MW01323098

VIARTIS

BARTOLOMEU DIAS

Ernst Georg Ravenstein
William Brooks Greenlee
Pero Vaz de Caminha

PUBLISHING DATA

TITLE : Bartolomeu Dias

AUTHORS : Ernst Georg Ravenstein, William Brooks Greenlee, Pero Vaz de Caminha

EDITORS : Keith Bridgeman, Tahira Arsham

ISBN : 978-1-906421-03-8

PUBLISHER : Viartis (http://viartis.net/publishers)

PUBLICATION DATE : 2010

PLACE OF PUBLICATION : England

LANGUAGE : English

FORMAT : Paperback

EDITION : First

TOPICS : Biography

LIBRARY CLASSIFICATION (Dewey decimal classification) : 920

SHORT DESCRIPTION : The biography of Bartolomeu Dias (c1450-1500), the Portuguese explorer who was the first European to sail around the southern tip of Africa.

LONG DESCRIPTION : The biography of Bartolomeu Dias (Bartholomeu Dias, Bartholomew Diaz) (c1450-1500), the Portuguese explorer who was the first European to sail around the southern tip of Africa. His discovery opened up the sea route around Africa to India and the rest of Asia. Bartolomeu Dias took part in more important voyages of discovery than any other explorer. He participated in voyages and discoveries along the west coast of Africa. He organised and accompanied Vasco da Gama's fleet on its voyage to India. Bartolomeu Dias finally captained a ship in the fleet of Pedro Cabral. It was the longest voyage in history up to that time, and one of the greatest and most influential voyages of discovery ever made. Unfortunately for Bartolomeu Dias, his return with that fleet to the site of his greatest discovery was to become the site of his greatest tragedy.

SIZE : 229 mm x 152 mm

PAGES : 126

INTRODUCTION

Bartolomeu Dias (Bartholomeu Dias, Bartholomew Diaz) was born in Portugal, probably in 1450.

Bartolomeu Dias and his crew were the first Europeans to sail around the southern tip of Africa. His discovery, which he described to his king in the presence of Christopher Columbus, opened up the sea route around Africa to India and the rest of Asia.

On his return, Bartolomeu Dias could have been considered the world's greatest discoverer. However, his discoveries did not cease there. Bartolomeu Dias was to take part in more important voyages of discovery than any other explorer.

Prior to his voyage to the southern tip of Africa, Bartolomeu Dias participated in voyages and discoveries along the west coast of Africa with Diogo da Azambuja.

After Christopher Columbus made voyages of discovery, Bartolomeu Dias organised and accompanied Vasco da Gama's fleet on its voyage to India. The voyage was only possible because of the earlier discovery of Bartolomeu Dias.

Because of the voyages of Bartolomeu Dias and Vasco da Gama, Portugal became mistress of the lucrative sea-route around Africa to India.

Bartolomeu Dias finally captained a ship in the fleet of Pedro Cabral, which was one of the largest fleets that had ever sailed the Atlantic. It included the discovery of Brazil as one of its achievements. It was the longest voyage in history up to that time, and one of the greatest and most influential voyages of discovery ever made.

On passing the site of his discovery of the southern route around Africa Bartolomeu Dias had taken "leave of it as from a beloved son whom he never expected to see again".

However, Bartolomeu Dias did return to the site of his greatest discovery, which unfortunately became the site of his greatest tragedy.

CONTENTS

PAGE	CONTENTS
1	Chapter 1 : The early life of Bartolomeu Dias
6	Chapter 2 : The first voyage of Diogo Cão (1482-1484)
13	Chapter 3 : The second voyage of Diogo Cão (1485-1486)
19	Chapter 4 : The voyage of Bartolomeu Dias (1487-1488)
31	Chapter 5 : The first voyage of Columbus (1492-1493)
36	Chapter 6 : The second voyage of Columbus (1493-1496)
40	Chapter 7 : The voyage of Vasco da Gama
43	Chapter 8 : The fleet of Vasco da Gama
51	Chapter 9 : The crew of Vasco da Gama
59	Chapter 10 : The voyage to Cape Verde Islands
62	Chapter 11 : The voyage to India
73	Chapter 12 : The voyage of Pedro Cabral
76	Chapter 13 : The fleet of Pedro Cabral
79	Chapter 14 : The crew of Pedro Cabral
90	Chapter 15 : Vasco da Gama's memorandum
94	Chapter 16 : The discovery of Brazil
98	Chapter 17 : Pero Vaz de Caminha's account of the discovery
118	Chapter 18 : The anonymous narrative

CONTENTS

PAGE	CONTENTS
124	Chapter 19 : The final voyage of Bartolomeu Dias
126	Bibliography

CHAPTER 1

THE EARLY LIFE OF BARTOLOMEU DIAS

Bartolomeu Dias is thought to have been born in Portugal in around 1450. Several Portuguese historians claim that he was a relative or descendant of João Dias who sailed around Cape Bojador in 1434, and of Dinis Dias who is said to have discovered the Cape Verde Islands in 1445. He was married and had two children : Simão Dias de Novais and António Dias de Novais. However, little is known for certain of the early life of Bartolomeu Dias.

There can be no doubt that Bartolomeu Dias was a seaman of considerable experience. It may have been our Bartolomeu Dias whom King João, in 1478, when still crown-prince, in consideration of 12,000 reis expended in the purchase of a slave, exonerated from payment of the usual royalty on the ivory bought on the Guinea coasts.

When Prince João, on August 26th 1481, ascended the throne of his father Afonso V to become King João II of Portugal, he found the royal treasury empty, and his ambitious nobles, jealous of their feudal privileges, ever ready to defy the authority of their king. However, King João II was strong and energetic where his father had been weak and vacillating, and cautious, where he had been rash and regardless of consequences. Wise measures of administration again filled the royal coffers, and a strong hand crushed the nascent conspiracy of the nobles.

He entered with zeal into the views of his predecessors and of his uncle Prince Henry. Before he came to the throne, a part of his revenues had been derived from the African trade, and the fisheries connected therewith, so that he had every inducement to prosecute its extension. With this view he not only ordered the completion of the Fort of Arguin, which had been commenced years before, but resolved on the construction of another, on a larger scale, at São Jorge da Mina.

The gold traffic had at first been carried on at a place called Saama, discovered in 1472, by João de Santarem and Pedro do Escover, in the service of Fernão Gomez, but São Jorge da Mina was now selected for its superior convenience. The gold mines around Elmina had been yielding profits to the Portuguese, but the whole method of trading was hazardous and chancy, particularly as the activities of unlicensed traders made life somewhat uncomfortable for the Portuguese.

King João II called his State Council and suggested the possibility of building a fort near the Elmina gold-mines. His councillors thought the idea preposterous, but King João overruled them and made plans for an expedition to be sent out to Elmina.

At the time, Elmina consisted of two townships - a section to the east of the River Benya, which was under the authority of the ruler of Feu State, and a section to the west of the river, which was under the Comanis or Komendas.

Diogo da Azambuja (1432-1518), a Portuguese noble and a distinguished Knight of the Royal Household, was chosen to command the expeditionary force. He was to build a fort at a suitable site near the Elmina gold-mines. That the fort might be constructed the most expeditiously, both for preventing objections and saving his people from exposure to the dangers of the climate, the King took the precaution to have the stones cut and fashioned in Portugal. With these, and bricks, and wood, and other needful materials, he loaded ten caravels and two smaller craft. He also sent out provisions sufficient for six hundred men, one hundred of who were officers to superintend the work.

Bartolomeu Dias accompanied Diogo da Azambuja on the voyage. The fleet set sail from Portugal on 11th December 1481. They took with them a great quantity of timber, hewn stone, line tiles, bricks, tools, nails, munitions, provisions, 100 craftsmen, including masons, and 500 soldiers.

After stopping to conclude a favourable treaty with Bezeguiche, the lord of the harbour and court that bore his name, they disembarked near Elmina on 19th January 1482 and decided to meet the local chief.

On the following morning, on 20th January 1482, Diogo da Azambuja seated himself on an imperial chair on Elmina beach, ordered the royal standard of Portugal to be hoisted and suspended from the bough of a lofty tree, at the foot of which they erected an altar. The whole company assisted at the first mass that was celebrated in Guinea, and prayed for the conversion of the natives from idolatry, and the perpetual prosperity of the church they intended to erect upon the spot.

By good luck they found there a small Portuguese vessel, the captain of which, João Bernardes, was engaged in traffic with the natives, and him they made interpreter between Kwame Ansa, the chief of Elmina (whom the Portuguese called Caramansa), and Diogo da Azambuja. An invitation was sent to the chief of Elmina.

Chief Kwame Ansa duly arrived with his retinue. The interview took place with the greatest ostentation possible on both sides, a kind of rivalry in which, as may be supposed, the African prince had a very sorry chance of producing any very imposing effect.

Azambuja appeared in a tunic of brocade, with a collar of gold and precious stones, and his captains were all in holiday attire, while Kwame Ansa, who was no less ambitious of making a good display, was habited, like the rest of his people, in the best vestments that nature had provided them. With their skins anointed and glistening till their native blackness was made blacker still, they considered their toilette perfect, although their only garment was an apron of monkeys' skin or palm leaves. To this extreme simplicity, however, Kwame Ansa himself was in so far an exception that his arms and legs were adorned with bracelets and rings of gold, and round his neck a collar from which hung small bells, and some sprigs of gold were twisted into his beard, so that the curls were straightened by the weight.

Diogo da Azambuja recounted the previous kindnesses shown by Kwame Ansa to Portuguese traders and requested permission to build a fort at Elmina. Diogo da Azambuja took pains to impress on Chief Kwame Ansa the might, power and dignity of Portugal, and the high position which he, Diogo da Azambuja, held there.

Kwame's answer represents the first recorded speech delivered by a

Gold Coast chief in reply to a representation made by an emissary from Europe, and it should help us to assess the mental condition of the unlettered Gold Coast African of centuries ago. Chief Kwame Ansa's words on 20th January 1482 were :

"I am not insensible to the high honour that your great master the chief of Portugal has this day conferred upon me. His friendships I have always endeavoured to merit by the strictness of my dealings with the Portuguese and by my constant exertions to procure an immediate lading for the vessels. But never until this day did I observe such a difference in the appearance of his subjects : they have hitherto been only meanly attired, were easily contented with the commodities they received, and so far from wishing to continue in this country, were never happy until they could complete their lading, and return. Now I remark a strange difference. A great number, richly dressed are anxious to be allowed to build houses, and to continue among us. Men of such eminence, conducted by a commander who from his own accounts seems to have descended from the God who made day and night, can never bring themselves to endure the hardships of this climate. Nor would they here be able to procure any of any luxuries that abound in their own country. The passions that are common to us all men will therefore inevitably bring on disputes; and it is far preferable that both our nations should continue on the same footing they have hitherto done, allowing your ships to come and go as usual; the desire of seeing each other occasionally will preserve peace between us. The Sea and Land being always neighbours are continually at variance, and contending who shall give way; the Sea with great violence attempting to subdue the Land, and the Land with equal obstinacy resolving to oppose the Sea."

Diogo da Azambuja then addressed the chieftain in the name of King João, commending to him the Christian religion, which if he would recognise and be baptised, the King would regard him as a brother, and make with him an alliance, offensive and defensive, against their common enemies, and enter into a treaty for the interchange of the products of their respective countries.

With this view he proposed, with the chieftain's permission, to found a permanent establishment in his country that should serve as a place of security against their enemies, as a refuge to the Portuguese who visited the coast, and also as a storehouse for their merchandise. Kwame Ansa, who was very shrewd for an African, after some hesitation, gave his consent.

One historian has remarked that this skilful evasion of the principle, if not the sole, object of the Portuguese mission considerably disconcerted the Portuguese. It required all the intellect of Diogo da Azambuja, backed by presents, promises and veiled threats, to induce Chief Kwame Ansa to give way. The Portuguese, however, knew that only the threats had induced Kwame Ansa to give way, and that if they could not make use of the present goodwill to build temporary defences, Kwame Ansa would soon seek help from his neighbours.

On the following day, Diogo da Azambuja put the work in hand, but no sooner was it commenced than the Africans showed signs of an intention to interrupt it.

Fortunately, mischief was prevented by Azambuja's learning that this arose from displeasure that the requisite presents had not as yet been offered to the chieftain. The oversight was soon remedied, and the work was set about with so much activity that in twenty days the fort was in a condition to repel an attack.

Diogo da Azambuja also built a church on the site, where on his arrival he had erected an altar. Both the church and the fort were dedicated to St. George. In the former, a daily mass was established in perpetuity for the soul of Prince Henry, and to the latter the King conceded the privileges of a municipality. Azambuja took up his abode there, with a garrison of sixty men, and sent back the rest of the men to Portugal with gold and slaves and other articles of merchandise.

By a charter of King João II, dated 17th March 1485, Diogo da Azambuja received in recognition of his great services in the wars, and especially in the construction of a fortress, the permission to add a castle to his arms in commemoration of the fact.

CHAPTER 2

THE FIRST VOYAGE OF DIOGO CÃO (1482-1484)

When King João had attended to what he conceived to be his more immediate duty as a king and ruler, he took up the long neglected work of his uncle Henry, for he was both "a good Catholic, anxious for the propagation of the faith, and a man of an inquiring spirit, desirous of investigating the secrets of nature".

Diogo Cão, whom the king selected to initiate this work of exploration, was a "man of the people". Genealogists have provided him with a noble pedigree, but he was in truth the descendant of one Pedro Affonso Cão, or Cam, who, in the days of King Diniz (1279-1324), had been one of the bailiffs of Villa Real in Traz os Montes, and of his wife, Briolanja da Nobrega. In the patent of nobility of 1484, by which the king "separated him from the common herd", the past services of the recipient of the honour are referred to, and also those of his valiant father, Badalhouce, and of his grandfather, Gonçalo Cão, who may have fought in the famous battle of Aljubarrota in 1385, when the Castillian pretender was routed, and the king, duly elected by the Cortes, got his own. Among the services rendered by Diogo Cão himself may be instanced the capture of three Spanish vessels on the Guinea coast in 1480.

Bartolomeu Dias may have been given the command of one of the vessels. However, none of the names of any of the officers have been placed on record. When we turn to ancient maps, we meet with a Rio do Infante, a Golfo de Alvaro Martins, a Cabo de Pero Dias, a Rio de Fernão Vaz, an Angra de João de Lisboa, an Angra de Ruy Pires, and a Serra de Corte Real. There can be no reasonable doubt that the names attached to these bays, capes, or rivers are those of persons who were with one or more of the expeditions engaged in the discovery of these coasts.

Of João Infante, Alvaro Martins, and Pero Dias, we know that they were part of the crew of Bartolomeu Dias, and may have previously been with Diogo Cão. João de Lisboa won great distinction in the course of time, and in 1525 was appointed Piloto mór of India. Unfortunately, he died the year after. Fernão Vaz may have been the pilot who, in 1486, witnessed the agreement between Fernão Dulmo and João Affonso do Estreito about the search for the Sette citades, who got into disgrace for poisoning his wife, and was himself poisoned by his mistress in 1502. The name of Corte Real we find on Behaim's globe only, and as the Corte Reaes of Terçeira were a family of seamen, it is quite possible that a member of it may have joined one of these expeditions, perhaps Gaspar, the alleged discoverer of "terra nova". It is, however, quite possible that Behaim merely intended to pay a compliment to a family with whom he was distantly related. Of Ruy Pires we know nothing.

Diogo Cão was the first to carry "padrões", or pillars of stone, on an exploring voyage. Up to his time the Portuguese had been content to erect perishable wooden crosses, or to carve inscriptions into trees, to mark the progress of their discoveries.

King João conceived the happy idea of introducing stone pillars, surmounted by a cross, and bearing, in addition to the royal arms, an inscription recording in Portuguese, and sometimes also in Latin, the date, the name of the king by whose order the voyage was made, and the name of the commander. The four padrões set up by Diogo Cão on his two voyages have been discovered in situ, and the inscription upon two of them (one for each voyage) were still legible centuries later, and had been deciphered.

During the first voyage two padrões were set up - one at the Congo mouth, the other on the Cabo do Lobo in latitude 13° 26' S.. The latter has been recovered intact. It consists of a shaft 1.69 metres high and 0.73 metres in circumference, surmounted by a cube 0.47 metres in height and 0.33 metres in breadth. Shaft and cube are cut out of a single block of lioz, a kind of limestone or coarse marble common in the environs of Lisbon. The cross disappeared, with the exception of a stump, from which it was seen that it was also of stone, and fixed by means of lead.

The arms of Portugal carved upon the face of the cube are those in use up to 1485, in which year João II, being then at Beja, caused the green cross of the Order of Avis, which had been improperly introduced by his grandfather, who had been master of that order, to be withdrawn and the position of the quinas, or five escutcheons, to be changed.

The inscription covers the three other sides of the cube. It is in Gothic letters and in Portuguese, and reads as follows :

"In the year 6681 of the World, and in that of 1482 since the birth of our Lord Jesus Christ, the most serene, most excellent and potent prince, King D. João I. of Portugal did order (mandou) this land to be discovered and these padrões to be set up by D° Cão, an esquire (escudeiro) of his household."

There is no inscription in Latin. As the year 6681 of Eusebius begins on September 1st 1481, we gather from this inscription that the order for the expedition was given between January and August, 1482. Of course the departure may have been delayed, but the delay can not have been a long one, as Diogo Cão was home again before April 1484.

Further light is thrown upon Diogo Cão's first voyage by a chart of Cristoforo Soligo, evidently drawn immediately after his return. Apart from this, we are dependent upon João de Barros and the chroniclers Ruy de Pina and Garcia de Resende. According to Ruy de Pina, Diogo Cão's first voyage was undertaken in 1485-1486, and according to Barros in 1484-1486. Both suppose Diogo Cão to have gone no further south than the Congo.

Diogo Cão may thus be supposed to have left Lisbon about the middle of 1482, possibly in June. He called at S. Jorge da Mina for supplies, and then made straight for Cabo de Lopo Gonçalves. His progress south along this coast was necessarily slow, for the current sets to the northward, the winds are southerly, and the surf is heavy. Only on rare occasion is the mariner favoured with a current setting to the south. He is dependent, therefore, for his progress upon a judicious use of land and sea breezes. Leaving behind him the Cabo de S. Catharina, with its "tree" marking the furthest point reached by the seamen employed by Fernão Gomez, the Cape of Pedro Dias, and the wooded hills of the

Holy Spirit, Diogo Cão seems to have made a first stay in a bay merely described as "Angra" on Soligo's chart, but named Gulf of Alvaro Martins on others, and later known as Mayumba Bay. He then left behind him a country of heavy rains and most luxuriant vegetation, and entered upon a region occasionally actually arid. Passing beneath the Paps of Bamba (os duos montes), and along a coast for the most part cliff bound, he entered the Bay of Loango, which must have taken his fancy, for he called it Praia Formosa de S. Domingos. It is possible that he arrived here on that saint's day, that is on August 4th. Passing thence to the south, and along fine red cliffs, Diogo Cão soon became aware that he was approaching a large river, for when still five leagues out at sea, as a legend on Soligo's chart tells us, he found the water fresh, which was by no means an exaggeration, for islands of floating vegetation coming out of the Congo have been encountered 100 miles from its mouth, and 9 miles to seaward the surface water is quite fresh.

Great must have been the astonishment, or terror of the natives, when for the first time they saw rising above the horizon the sails of a white man's vessel, and beheld the bleached faces of its inmates. Diogo Cão sailed up the river for a short distance, and at once entered into friendly relations with the natives. Physically, they resembled the inhabitants of Guinea, but the interpreters whom Diogo Cão had with him failed to make themselves understood. The natives came freely on board to barter cloth in exchange for ivory, and gave their visitors to understand by signs that far in the interior there lived a powerful king.

Diogo Cão at once despatched some Christian natives to this king as his ambassadors. They were, as a matter of course, the bearers of suitable presents, and were instructed to assure the king of the friendly intentions of his visitors from Portugal, and of their desire to trade. The native guides promised to bring these messengers back within a certain number of days.

Before leaving the Congo for the south, Diogo Cão set up the first of his padrões, emphatically called "the first" on Canerio's chart. It stood on Shark point, and not on Padron point of our chart, and was dedicated to S. Jorge, a saint for whom King João felt a "singular devotion". We learn from Fathers Cavazzi and Merolla that the Dutch, when they occupied the Congo in 1642, wantonly destroyed this memorial of Portuguese enterprise. Merolla, who saw the fragments in 1682, was able to trace the royal arms and an inscription, of which unfortunately he made no copy. A tall wooden cross was subsequently erected on the spot where the pillar stood, and an oratory built near it where masses might be said. The fragments of the padrão were appropriated by native priests, who looked upon them as most potent fetishes. Sr. Sori saw them in 1859. Burton visited the locality in 1863, and Baron Schwerin, guided by Sr. F.J.de França, did so in April 1887. The baron caused the bandages in which the fragments were wrapped up to be removed, in honour of which event the Massebi, a Portuguese gunboat, fired a salute. Mr. Dennett examined the fragments in May of the same year. There were two large pieces and two ball-shaped pieces, 7 and 9 inches in diameter, lying at their foot. The material was a coarse white marble. The two larger fragments were later put in the museum of the Lisbon Geographical Society. The very hideous monument set up in 1859 by Sr. Sori was fortunately washed away by the sea in 1864. It was replaced, in 1892, by a memorial of better design, but still vastly inferior to the original padrão.

It is to be presumed that Diogo Cão, when, after a delay which may well have extended over several months, left for the south, was able to gain a fairly complete knowledge of the coast, for his progress must have been slow. We know from Soligo's chart that he discovered a river that he named after Fernão Vaz, the river Dande, which is shown on Soligo's chart, and also the low sandy Ilhas das Cabras (Goat islands), off the modern city of Luanda.

Making a long stretch from the coast, Cão never noticed the most important river along the coast, the Kwanza, although its clayey waters discoloured the sea for 10 or 15 miles. It is curious that none of his immediate successors should have been more fortunate. The river is not mentioned in Pacheco's "Esmeraldo", and is apparently shown for the first time on a chart of P.Reinel, who already knew its native name.

A remarkable headland, which from some bearings appears as a double peak, was appropriately named by Diogo Cão, "A terra da duas Pontas". It is now known as the "Morro" or hill of Old Benguella.

Further south, Diogo Cão seems to have examined the mouth of the Catumbela, for Rio do Paul (river of the swamp) is a very appropriate name for a river, which after the rains in March and April, overflows its banks and converts a great extent of country into a swamp or marsh. As Cão called the bay to the south Angra de St. Maria, he may have been in the vicinity of this river on Lady Day, on March 25th 1483.

The bold granitic cliffs immediately to the south of Ponta Choca (13° 17' S.) became known as Castello d'Alter pedroso. About 10 miles beyond, on a low point, which he called "Cabo do lobo" (seal point), Cão erected his second padrão, which was dedicated to St. Augustin, from which it must not be inferred that it was erected on August 28th, as these dedications were made in Portugal. At the back of this cape, later known as S.Maria (13° 26' S.), rose a Monte negro (black mountain). Pacheco tells us that it was called Ponta negra, or Preta (both meaning black point), because of a "black" trump, "manilha negra", which was played here in a game of manille. This padrão has already been described by us. It was the second and last set up during this voyage, and Soligo's chart correctly describes it as "o ultimo padrão".

When Diogo Cão returned to the Congo, he was annoyed to find that the messengers whom he had dispatched to the king had not yet returned, although they had been absent double the time expected.

Diogo Cão, who was naturally anxious to return home with a report of his discovery of what seemed a powerful kingdom, therefore seized four native visitors to his ship as hostages, giving their friends to understand that after the lapse of fifteen months he would bring them back and exchange them for his own men, who were still with the king. These latter, we learn from Ruy de Pina, had been made much of, but when the king heard of Diogo Cão's highhanded proceedings, he refused to admit them any longer to his presence, and threatened to kill them, should his own people not be restored in time.

Among Diogo Cão's hostages was one Caçuto, a "nobleman" in his own country, and a man of some intelligence, who seems to have picked up Portuguese rapidly. King João was much pleased with this man, and the information he was able to give. He, as well as his companions, were treated with much distinction, and dressed in fine cloth and silk.

Diogo Cão came back to Lisbon probably in the beginning of 1484, and certainly before April of that year. The king, first of all, made him a "cavalleiro" of his household. The king then, on April 8th 1484, "in consideration of the services rendered in the course of a voyage of discovery to Guinea, from which he had now returned", granted him an annuity of ten thousand reals, to be continued to one surviving son. A few days afterwards, on April 14th he separated his "cavalier" from the common herd and made him noble, and gave him a coat-of-arms charged with the two padrões, which he had erected on the coast of Africa.

CHAPTER 3

THE SECOND VOYAGE OF DIOGO CÃO (1485-1486)

The materials for writing a history of Diogo Cão's second expedition are even less complete than those available for the first. There are the padrão of Cape Cross with its inscription, an important legend on the chart of Martellus Germanus, and the narrative of Martin Behaim, who claims to have commanded one of the vessels. Apart from these, we are dependent upon the accounts given by Ruy de Pina and João de Barros, for none of the later historians seem to have had access to original sources.

The narrative of Behaim, as gathered from the legends on his famous globe, and a paragraph in Schedel's "Liber Chronicorum", printed at Nürnberg in 1493, during Behaim's presence in that town, is written as follows :

"In 1484 King João sent two vessels to the south, one being commanded by D. Cão, the other by Martin Behaim. They carried, in addition to goods for barter, eighteen horses with splendid harness, intended as presents for Moorish (i.e. Negro) kings. They traded with the Jolof and on the Gambia, visited King Furfur's land [Benin], 1200 German leagues from Lisbon, where the Portugal pepper grows, and came to a country where they found cinnamon. They also discovered Prince's island, S.Thomé and Martin [Behaim's!] islands (i.e. Annobom). On January 18 they set up a column on Monte Negro (Cão's third pillar in 15° 40'). Having sailed 2300 leagues, they set up another pillar on Cape Ledo. They were again with their king after an absence of 19 (16 or 26) months, having lost many men from the heat, and bringing pepper, grains of paradise, and many other things in proof of the discoveries they had made."

We have elsewhere considered the trustworthiness of this account of Diogo Cão's expedition, and arrived at the opinion that Behaim did not accompany Diogo Cão, but may have been on the Guinea coast with an expedition such as that of João Affonso d'Aveiro.

Far more useful for our purpose is the pillar that formerly stood on Cape Cross, and which Captain Becker of the Falke carried off to Kiel in 1893. Dr. Scheppig has fully described the pillar.

The shaft is 1.84 m. long, and has a circumference at the bottom of 0.93 m.. It tapers slightly towards the top, and is surmounted by a cube 0.43 m. high, 0.45 m. broad, and 0.26 m. thick. The whole is hewn out of a single block of marble. The cross, also of marble, was fixed by means of lead. The arms carved on the face of the cube are those adopted by João II in 1485. There are two inscriptions in Gothic characters, the one in Portuguese, the other in Latin.

The Portuguese inscription says : "In the year bjMbjclxxxb (6685) of the creation of the world, and of Christ llllclxxxb (485), the excellent, illustrious King D.João II of Portugal did direct this land to be discovered, and this padrão to be set up by D. Cão, a cavalleiro (knight) of his household".

The Latin inscription reads as follows : "There had elapsed 6684 (5?) years since the creation of the world, and 148- since the birth of Christ, when the most excellent and most, serene King, D.João II of Portugal... ordered this column to be set up by his knight (militem) Iacobus Canus (i.e. Diogo Cão)."

Dr. Scheppig observes that the dates in this Latin inscription are both written in Arabic characters, "which, owing to their novel form, were still sources of frequent error and confusion", and that the fourth cipher in 6684 is certainly of abnormal shape, and may perhaps be meant for a 5, in which case both inscriptions would agree. As to "1485 " no doubt whatever arises.

As the year 6685 of the Eusebian era begins on September 1st 1485, Diogo Cão must have departed after that day, and before the close of the year. As he had returned from his first voyage before April 1484, his departure must have been delayed for reasons not known to us. Perhaps it was owing to the opposition of the Royal Councillors to further expeditions, perhaps a desire that the contemplated change of arms might be recorded on the padrões to be sent with the explorer.

During this voyage Diogo Cão seems to have commanded a fleet - at least, so we are told by Ruy de Pina, Garcia de Resende, and Martellus Germanus. He took with him, as a matter of course, the four men whom he had so unceremoniously carried off. These had been well treated in Portugal, and were the bearers of rich presents to their king, whom they were to invite to throw aside his idols and fetishes and embrace the only saving faith.

It may be presumed that Diogo Cão, in the course of this second voyage, gained a fuller knowledge of the coast first discovered by him to the north of the Congo. He may thus have visited and named the bay called Golfo do Judeu, the Jews' bay, of old maps, either because there was a Jew on board his vessel, or, what is less likely, because he was struck with the Jewish physiognomy of some of the natives, who are absurdly supposed to keep the Jewish sabbath, when in reality they have fetishes and Casas da tinta like their neighbours. He may also have entered the fine Golfo das almadias (Kabinda Bay), still famous for its boats, as it appears to have been in the days of the early Portuguese.

There was great rejoicing when Diogo Cão entered the Congo, and it became known that the hostages whom he had carried off were on board his ship. He at once sent one of these men to the Mani Congo, to announce his arrival, and to beg that his own people should he sent down to the coast, when the other three would be released. When the man came back, Diogo Cão sent a present to the king, and let him know that he was about to follow the coast to the south, but that on his return he would seek speech with him, and hand over the presents with which he had been entrusted.

Passing southward along the coast, Diogo Cão landed several times for the purpose of carrying off natives who were to be taught Portuguese, so that on future occasions they might act as interpreters. Near Cape St. Bras he saw native fishgarths, and hence called that double of the Bill of Portland Ponta das Cambóas.

When about 160 miles beyond the second padrão set up by him in 1483, he reached a second Monte Negro, a remarkable headland in 15°

41' S., rising like an island to a height of 200 feet, and presenting a rugged black face towards the sea, and upon this he set up a padrão. The bay to the south he named Angra das Aldeas, because of two poor fishing villages. In design and size this padrão resembles that of Cape Cross. A trace of the royal crown is still visible, but time has obliterated the inscriptions.

The aspect of the country had gradually grown poorer and poorer, until barren sandhills and arid rocks were all that could be seen from the sea, except at a few openings where streams or rivers had given birth to vegetation and verdure (praias verdes). Lofty mountains now and then were visible far inland. Passing along such a coast of low sandhills and white cliffs, Diogo Cão came past the broad Golfo da Baleia (Whale bay), separated by a "sleeve of sand " (Manga de Areia) front the open sea.

He must have noticed the low black rocks with yellow specks, first known as Cabo preto, and ultimately reached a truncated cone of red sandstone, in 21° 50', upon which he set up the last of his padrões, already fully described by us. This was one of the Cabos do padrão of old charts, and is now known as Cape Cross. To the south of it, on Martellus Germanus's chart, we notice a Praia das Sardinhas (Sardine shore), now known as Sierra bay, and a Serra parda, which may safely be identified with the dark and rocky cliff now known as Cabo dos Farilhões (22° 9' S.), surmounted by a sandy dune, and rising inland into peaks.

This cape, 430 leagues, or 1450 sea miles, to the south of Cape Catharina, is the furthest point reached by Diogo Cão, and if a legend on the chart of Henricus Martellus Germanus may be accepted, he died there. This legend is to the following effect:

"This mountain, called the Black mountain [i.e. Monte negro, in 15° 41'] was reached by the fleet of [João] the second King of Portugal, which fleet was commanded by Diegus Canus, who, in memory of this fact, set up a marble column, with the emblem of the cross, and proceeded onwards as far as the Serra parda, which is distant 1000 miles from the Black mountain, and here he died".

A "parecer" or opinion, drawn up by the Spanish astronomers and pilots who attended the congress of Badajoz in 1525, and signed by Hernan Colon, Juan Sebastian del Cano, and others, goes far to confirm this legend, for it tells us that Cão, in the course of his second voyage, discovered the coast from Montenegro as far as the Sierra Parda, where he died, a distance of 200 leagues (680 sea miles).

The distance between Cape Negro and Sierra Parda actually amounts to 435 sea miles (139 leagues, or 556 Italian or Roman miles), but if we assume the Mediterranean on the chart of Germanus (which has no scale) to measure 3000 Italian miles in length, as usually adopted, then the distance separating Montenegro from Sierra Parda on that chart would equal 1000 of these miles.

Of course, if Diogo Cão died near his last padrão, we are compelled to reject the account given by Ruy de Pina and Barros of the final stages of his expedition, and what is generally accepted. According to these historians, Diogo Cão returned to the Congo, had an interview with the Mani Congo, who expressed a desire for priests to convert his people, masons and carpenters to build churches and houses, labourers to break in oxen, and women to make bread, so that his kingdom might in every respect become like Portugal. He sent Caçuto, one of Diogo Cão's hostages, as ambassador to Portugal, and with him the sons of several of his courtiers, desiring that they should be taught to read and write and made Christians. At the same time he sent a present of ivory and palm cloth, the most valuable products of his kingdom.

Now, we have good reason to believe that Caçuto was received by the king in the beginning of 1489, the king being then at Beja, where he and his companions were baptised with much solemnity, the king himself, his queen, and gentlemen of title acting as sponsors. We know further, that Caçuto, henceforth known as D.João da Silva, was sent back to Congo with D.Gonçalo de Sousa, King João's ambassador, in December 1490. Barros says that this happened two years after he had been baptised.

It might reasonably be concluded, from these dates, that Caçuto arrived in Portugal in December 1488, was baptised at Beja in January 1489, and again left for Congo, after a stay of two years, in December 1490. However, if this be so, he cannot have come with Diogo Cão, for Cão, or his ships, must have been back before August, 1487, in which month Bartolomeu Dias started on his voyage, taking with him the people whom Cão had kidnapped. However, in all probability Diogo Cão's ships came home even earlier, say in September, 1486, for on October 10th of that year, Bartolomeu Dias seems already to have been appointed to the command of the expedition which was to make him famous for all time.

Indeed, we are inclined to think that after Diogo Cão's death, his vessels returned straight home, and if they did so, and the Eusebian era is stated quite correctly on the padrão of Cape Cross, they can have been away at the outside for thirteen months, that is, from September 1485, to September 1486 - not a long period, but amply sufficient for a voyage to Cape Cross and back, and a stay of several months on the Congo river.

CHAPTER 4

THE VOYAGE OF BARTOLOMEU DIAS (1487-1488)

No sooner had Diogo Cão's vessels returned to the River Tagus than King João, whose curiosity had been excited by the reports about the supposed Prester John, brought home by d'Aveiro, determined to fit out another expedition to go in quest of him by doubling Africa. Friar Antonio of Lisbon and Pero of Montaroyo had already been dispatched on the same errand by way of Jerusalem and Egypt.

The command of this expedition was conferred upon Bartolomeu Dias de Novaes, a cavalier of the king's household, who, if we may trust Fernão Lopez de Castanheda, held at the time the appointment of superintendent of the royal warehouses (almoxarife dos amazons).

In writing the accounts of the voyages of Diogo Cão and Bartolomeu Dias, we have largely profited by a few contemporary maps. These maps, unfortunately, are on a very small scale. This compelled their compilers to confine themselves to a selection among the place-names that they found upon the sailing charts at their disposal, and this selection may not in all cases have been a judicious one.

The appointment of Bartolomeu Dias seems to have been made in October 1486. For on 10th October 1486, King João, "in consideration of services that he hoped to receive," conferred upon Bartolomeu Dias, the "patron" of the S.Christovão, a royal vessel, an annuity of 6000 reis. We shall see presently that ten months were allowed to elapse before the expedition actually left the Tagus.

The account that João de Barros has transmitted to us of the remarkable expedition that resulted in the discovery of the Cape of Good Hope is fragmentary, and on some points undoubtedly erroneous.

Until now, no official report of the expedition has been discovered. However, there are a few incidental references to it that enable us to amplify, and in some measure to correct, the version put forward by the great Portuguese historian.

Most important among these independent witnesses is a marginal note on folio 13 of a copy of Pierre d'Ailly's "Imago mundi", which was the property of Christopher Columbus. This reads as follows :

"Note, that in December of this year, 1488, there landed at Lisbon Bartolomeu Didacus [Dias], the commander of three caravels, whom the King of Portugal had sent to Guinea to seek out the land, and who reported that he had sailed 600 leagues beyond the furthest reached hitherto, that is, 450 leagues to the south and then 150 leagues to the north, as far as a cape named by him the Cape of Good Hope, which cape we judge to be in Agisimba, its latitude, as determined by the astrolabe, being 45° S., and its distance from Lisbon 3100 leagues. This voyage he [Dias] had depicted and described from league to league upon a chart, so that he might show it to the king; at all of which I was present (in quibus omnibus interfui)."

The same voyage is referred to in a second "note" discovered in the margin of the "Historia rerum ubique gestarum" of Pope Pius II, printed at Venice. From this second note we learn that "one of the captains whom the most serene King of Portugal sent forth to seek out the land in Guinea brought back word in 1488 that he had sailed 45° beyond the equinoctial line."

Las Casas, in "Historia de las Indias", assumed these notes to have been written by Bartholomew Columbus, whom, as the result of a misconception of the meaning of the concluding words of the note, he supposed to have taken part in this voyage.

These assumptions, are inadmissible, for as early as February 10th 1488, Bartholomew had completed in London a map of the world for Henry VII. If we remember that Bartholomew was detained by pirates for several weeks before be reached England, he must have left Lisbon towards the end of 1487. He did not return to that place until many years afterwards.

On the other hand, the note is unhesitatingly recognised as being in the handwriting of Christopher Columbus by such competent authorities as Varnhagen, d'Avezac, H.Harrisse, Asensio, and Cesare de Lollis.

If Christopher Columbus is the author of these notes, they must have been written in 1488. For it was on March 28th 1488, that King Manuel, in response to an application, cordially invited his "especial friend", Christopher Columbus, to come to Lisbon, promising him protection against all criminal and civil proceedings that might be taken against him. Such a promise was needed, for Columbus, in 1480, stole away from Lisbon without paying 220 ducats, which he owed to certain of his creditors. Columbus, when he received this royal invitation, was at Seville, where his son Ferdinand was born unto him on September 28th 1488. If he left Seville soon afterwards, he may certainly have been present on the memorable occasion, in December 1488, when Bartolomeu Dias rendered an account to the king of the results of his hope inspiring voyage.

If then, Bartolomeu Dias returned in December 1488, after an absence (according to De Barros) of sixteen months and seventeen days, he must have started towards the end of July or in the beginning of August 1487. If the Bartolomeu Dias referred to in the royal rescript of October 10th 1486 is the discoverer of the Cape, which hardly admits of a doubt, he cannot have started in July 1486, as is usually assumed. He cannot have been in Lisbon in December 1487.

This date (namely 1488) is further confirmed by Duarte Pacheco Pereira, the "Achilles Lusitano" of Camoens, for in his "Esmeraldo de Situ Orbis", written soon after 1505, but only published in 1892, we are told that the Cape was discovered in 1488. Pacheco is a very competent witness, for Dias, on his homeward voyage, met him at the Ilha do Principe.

Turning back now to Colon's "note," we find that Bartolomeu Dias is supposed to have sailed 450 legoas, or 25.3° to the south of Diogo Cão's furthest; and as Cape Cross actually stands in latitude 21.8°, this would have brought him to latitude 47.1° S. A return voyage of 150 legoas, or 8.5° to the northwards, would have reduced his latitude to 38.6° S. However, if Colon assumed Cape Cross to be in latitude 19° S., as on Dr. Hamy's and the Cantino charts, then the highest latitude reached would have been 44.3° S. We are justified in concluding from this that Colon's 45° does not refer to the Cape, but to the highest latitude reached. As to the 3100 leagues, the supposed distance from Lisbon, we have evidently to deal with a slip of the pen, for the distance to the Cape, following the coast, is only 6000 miles, or 100°.

A further statement respecting the date of the discovery of the Cape appears in the Parecer, or "opinion" of the Spanish astronomers and pilots already referred to. They say, "And beyond this [the Sierra Parda, where Cão died], Bartolomé Diaz, in the year 1488, discovered as far as the Cabo d'El-Rei, a distance of 350 leagues, and thence to the Cabo de boa Esperança, 250 leagues; and thence D.Vasco da Gama discovered 600 leagues...".

The distances given are exaggerations, for it seems to have been the object of these "experts" to push India and the Moluccas as far to the east as possible, so that the latter might fall within the Spanish sphere : the coastline actually discovered by Bartolomeu Dias measures less than 380 leagues. The nomenclature given is curious, for the designation of Cabo d'El-Rei is bestowed upon the Cape of Good Hope, and the latter name, not inappropriately, transferred to the furthest point reached by Bartolomeu Dias. We have not come across a single chart or document bearing out this nomenclature.

There remain to be noticed two references to the expedition of Bartolomeu Dias in the "Roteiro" describing Vasco da Gama's first voyage, for which we are indebted to Pero d'Alemquer, the pilot of Dias' flagship; and the statement of John of Empoli, the supercargo of one of the vessels of Affonso de Albuquerque's fleet (1503), that the Bahia dos Vaqueiros of Dias was renamed Bahia de St. Braz, because it was discovered on the day of that saint.

Bartolomeu Dias is supposed to have erected three padrões, but only one of these has until now been discovered. As the inscription upon it is no longer legible, it furnishes no evidence of the date of the voyage.

This pillar stood on Dias Point, South of Angra pequena, or Lüderitz Bay. Sir Home Popham saw it in 1786, but even then the inscription could no longer be deciphered. Captain Vidal, in 1823, found the pillar in fragments. The shaft, of marble, rose originally about 6 feet above the ground, and was buried to a depth of 21 inches. It was surmounted by a stone cross 16 inches high. In 1856 Captain Carrew brought three fragments to Cape Town, two of which, in 1865, were handed over to Chevalier du Prat, and were then sent to Lisbon, whilst the third, 22 inches high, 8 inches broad, and 5½ inches thick, remained in the Cape museum.

The "pillars" carried away by Bartolomeu Dias seem to have resembled those entrusted to his predecessor, Diogo Cão, except that, in addition to the royal arms, there was carved upon them a pelican, the device that King João had assumed when a prince, together with the motto, "Por tua ley e por tua grey". Such, at least, would appear to have been the case, to judge from the description of a series of pictures, illustrating the discovery of India, which were to have been painted by order of King Manuel.

Apart from what can be gathered from the above, and from a few early maps, we are dependent upon De Barros for what we know concerning the voyage of Bartolomeu Dias. Other historians have either slavishly copied him, or they adduce no fresh information. Strange to say, Ruy de Pina and Garcia de Resende, the chroniclers of King João II, although they refer casually to the discovery of the Cape of Good Hope, do not once mention the name of Bartolomeu Dias. As for Correa, the author of the "Lendas da India", he may safely be discarded. Correa does not mention the name of Dias.

Bartolomeu Dias was given the command of two ships, of fifty tons each, and of a store-vessel. His flagship, we think, must have been a Christovão, perhaps the very vessel that he commanded before his departure in 1486, and again in 1490-1495; or possibly a new vessel bearing the familiar name.

His chief pilot was Pero d'Alemquer, an experienced seaman, who subsequently served under Vasco da Gama. The master's name was Leitão. The second ship, the St. Pantaleão, had for its captain João Infante, a cavalier of the king's household, with whom were Alvaro Martins as pilot, and João Grego as master. The store-vessel was placed in charge of Pero Dias, a brother of Bartolomeu. João de Santiago was pilot, João Alves master, and Fernão Colaço, of Lumiar, clerk.

There were on board two African natives whom Diogo Cão had kidnapped, and also four African women from the Guinea coast. Strict orders had been given, not only to avoid every conflict with the natives, but also to gain their confidence using gifts. The four Guinea women were to be landed at various places, handsomely dressed, and furnished with samples of gold, silver, and spices, which the Portuguese were in quest of. These they were to exhibit wherever they went, proclaiming, at the same time, the greatness and munificence of the King of Portugal, and the ardent desire that possessed him to communicate with Prester John. Women were selected for this duty, as they would be respected even in the midst of tribal wars.

Bartolomeu Dias, we have no doubt, was furnished with a copy of the chart compiled by D. Diogo Ortiz de Vilhegas, of Calçadilha, Dr. Rodrigo, of Pedras Negras (the king's physician), and Master Moses, a Jew; which had been given in May 1487, to Pero de Covilhã.

The expedition left Lisbon at the end of July or in the beginning of August 1487, and sailed directly for the Congo, beyond which the coast was examined with attention, capes and bays being named either after saints, on account of striking physical features, or in connection with some occurrence in the course of the voyage.

On reaching the Angra do Salto, which we conceive to be identical to the Golfo das Aldeas, which was later known as Port Alexander, the two Africans carried off by Diogo Cão were restored to their friends.

It is just possible that the store-vessel was left in this safe and commodious harbour, where fish abounded, good water was plentiful, and natives with herds of sheep and bullocks were within reach. Cosa's Golfo do Saco and the Golfo do Salto of the Cantino map refer, no doubt, to the same locality. Cosa places the name near where Port Alexander should be, whilst Cantino locates it on a barren coast between that port and Great Fish bay, which was not frequented even by fishermen.

Struggling against south-westerly winds and a current setting to the north, Bartolomeu Dias passed the last pillar set up by Digo Cão. We have absolutely no direct information about Dias' proceedings from the time he left Angra do Salto to his arrival at Cabo da Volta, where he set up his first pillar. He may have named the country to the south in honour of S. Barbara, whose day is December 4th, and entered on December 8th the Golfo de S.Maria da Conceição, our modern Walvis Bay. Here he seems to have tarried, for, taking the saints' names bestowed along this coast for our guide, the next locality named by him must have been the Golfo de S.Thomé (December 21st), only 145 miles beyond. It cannot have taken a fortnight to make so short a run. It was probably here that the first of the African woman was landed.

We may then suppose Bartolomeu Dias to have sailed southward along the desolate coast of sandhills, where he possibly experienced the hot blasts of an easterly wind, and hence bestowed upon this forbidding region the appropriate name of "Areias gordas", that is, "hell." The gulf of S.Thomé was probably named on December 21st.

A few days afterwards, Bartolomeu Dias arrived at the Cabo da Volta and the Serra Parda, where he erected the padrão dedicated to Santiago, fragments of which have been recovered, and have already been referred to. He also landed here the second of the African women, probably leaving her with natives who had come down to the shore to fish.

On Cantino's chart, the deep bay to the east of this cape, our modern Angra Pequena, is called Golfo de S.Christovão. This, it appears, was the name originally bestowed upon a bay that subsequently became known as Angra or Golfo das Voltas - the "bay of tacks".

It is not probable that Bartolomeu Dias remained long in this bay. De Barros tells us that he stood off and on for five days, when there arose a strong wind, which compelled him to reduce his sails, and before which he ran south for thirteen days. This statement we are not prepared to accept, for northerly winds are exceedingly rare along this coast, and the squalls from the north-north-east or north-north-west, which are experienced occasionally, are never of long duration. However, when

Bartolomeu Dias reached a higher latitude on the south-east edge of the Agulhas bank, and came under the influence of the "roaring forties", it is very likely that he met with gales and a heavy sea. Considering the small size of the vessels, his men are to be excused if they stood in "mortal fear", and they naturally suffered from the cold, for in these latitudes they experienced a mean temperature of 10° or less, which is hard to bear for men fresh from a tropical climate.

During the first period of this long stretch to the south, Bartolomeu Dias maybe supposed to have kept within sight of the coast. He may thus have named the Golfo do S.Estevão, subsequently Elizabeth Bay, on December 26th, and the Terra da Silvestre on December 31st.

He may even have heard the roar of the rollers thundering upon the shore of the Terra dos Bramidos, and gained a view of the lofty Serra dos Reis on January 6th 1488. However, beyond these he lost sight of the land, and when passing St. Helena Bay, as we learn from Pero d'Alemquer, he was far out at sea.

When the storm subsided, Bartolomeu Dias stood east, and having failed, in the course of several days, to meet with land, he turned his prow to the northward. Sailing in that direction for 150 leagues, he saw lofty mountains rising before him, and came to anchor in a bay that he called Bahia dos Vaqueiros (Cowherd's bay). This happened on 3rd February 1488, and as this day is dedicated to St. Blaise, the bay, so we are told by John of Empoli, was renamed Bahia de S.Braz. It is now called Mossel Bay, and is in South Africa.

We learn from Pero d'Alemquer that the natives refused the presents that were offered them, and when Bartolomeu Dias landed to take in water from a well close to the beach, he was pelted with stones. One of the natives was killed with an arrow from a crossbow. They then retreated inland with their cattle.

During his onward course Bartolomeu Dias had to struggle against the Agulhas current, as also against the prevailing south-easterly winds. He may, however, have taken advantage of an inshore counter-current, setting eastward, as also of occasional westerly winds. At all events, he made his way along a coast bounded by lofty mountains. Rounding Cabo do Recife (Cape Recife), be entered a vast bay, which was called Bahia da Roca (Rock Bay), but which was subsequently known as Algoa Bay. Within it lay a group of rocky islets - the Ilhéos da Cruz.

Duarte Pacheco says that the tallest of this group is also known as Penedo das Fontes (Fountain rock) because of two springs that rise upon it, and that Bartolomeu Dias erected a pillar there, visible out at sea. As a matter of fact, the largest of these islets is nearly all bare rock, and there are no springs. Nor does it appear, if we may accept the results of M. de Mosquito Perestrello's careful survey of this coast in 1575, that a stone pillar was ever set up on this islet, notwithstanding the name it bears.

It is, however, possible that Bartolomeu Dias erected a wooden cross upon it, all traces of which had disappeared when Perestrello examined the coast. The islet may have been named because it was discovered on the day of the invention of the cross, which was May 2nd. In that case Dias must have spent three months in making good the 200 miles that separate it from Mossel Bay, which is difficult to believe.

Having left here the last of his African women (one of them having died during the voyage) in the company of two women who were gathering shellfish along the beach, Bartolomeu Dias continued his voyage. He sailed past the Ilhéos Chãos (Low islands), and about 12 miles beyond them, at or near a sandy cliff, still known as Cape Padrone, he set up a padrão dedicated to S.Gregorio. It is quite possible that this pillar was erected on St. Gregory's day (March 12th), though as a rule these dedications seem to have been made at home. Perestrello says the pillar stood on an islet at the foot of this cliff, the only islet between the Ilhéos Chãos and the Rio do Infante, but only sunken rocks are met there now, and the islet may have been destroyed by the force of the breakers. We do not gather from Perestrello's account that he himself saw the pillar.

It was probably about this time, when the coast was actually seen to stretch away towards the north-east, in the desired direction, that the ship's companies began to murmur about the hardships to which they were being exposed. Bartolomeu Dias, whose Regimento, or instructions, directed him to consult his officers on all occasions of importance, therefore invited them to land with him, together with a few leading seamen.

The result of this council was a decision in favour of a return, and a document to that effect was signed by all present. Bartolomeu Dias, however, persuaded his followers to go on eastward for two or three days longer. He promised that, unless something happened within that period to induce them to change their minds, he would accede to their wishes.

He was thus able to pass the remarkable rock identified by Perestrello with the Penedo das Fontes, where the dammed-up waters of a small stream soak through the beach ridge. This, we have no doubt, is Ship rock. The Rio do Infante (Great Fish river) lies only about 16 miles beyond. It was thus named because the captain of the Pantaleão was the first to land at its mouth. Here Bartolomeu Dias turned back. Galvão says that he saw "the land of India, but, like Moses and the promised land, he did not enter it".

On passing his padrão, "he took leave of it as from a beloved son whom he never expected to see again". His forebodings tragically proved to be true.

During his homeward voyage, Bartolomeu Dias was favoured by winds and currents. It is almost certain that he named the Cabo do Infante, and probably that he dedicated the southernmost cape of all Africa to St. Brendan, an apocryphal Irishman, whose day is May 16th.

It was soon after this that he beheld, for the first time, and coming from the east, the remarkable group of mountains, broken land, or "terra fragosa", as the ancient maps have it that fill Cape Peninsula. The southern extremity of which, if we may believe Barros, he named Cabo Tormentoso, in memory of the storms that he had experienced, a name which the king, whose hopes of reaching India by an ocean route seemed about to be realised, changed into Cabo da boa Esperança - the Cape of Good Hope.

It seems that this is one of those pretty legends frequently associated with great events. Duarte Pacheco, a contemporary, distinctly tells us that it was Bartolomeu Dias who gave the Cape its present name. Christopher Columbus, who was present when Bartolomeu Dias made his report to the king, says the same. Barros, indeed, seems alone to be responsible for this legend, for if Camoens speaks of a Cabo Tormentorio", we must remember that he lived through the terrible tempest that overwhelmed a part of Cabral's fleet. This was during the season of storms, in winter. Bartolomeu Dias, who spent several months on the south coast, may of course have met with gales, which would justify an appellation such as "Cape of Storms".

Still, on his homeward voyage, when alone he was in the immediate vicinity of the Cape, he seems to have been fortunate, for Pero d'Alemquer, his pilot, informs us that he left the Cape on a morning with a stern wind, which rapidly carried him northward. Before leaving the Cape, Bartolomeu Dias erected the last of his padrões, which was dedicated to St. Philip. The site of this pillar is absolutely unknown.

After an absence of nine months, Bartolomeu Dias rejoined his storeship. He found that six men had been killed in a trade dispute with the natives, and Fernão Colaço, one of the three survivors, died of joy on beholding his comrades. The vessel, being worm-eaten, was burnt after the provisions had been taken out of her.

It was now about the middle of August 1488, if we assume Bartolomeu Dias to have parted from his store-vessel about the middle of November. There thus remained four months for Bartolomeu Dias to make his way to Lisbon. Of what he did during these four months we know very little.

We do not even know whether he called at the Congo. We know, however, that he touched at the Ilha do Principe, where he met Duarte Pacheco with part of his shipwrecked crew, all of whom he took on board, and also that he then touched at Rio do Resgate - trade river - where he seems to have purchased some slaves, "so as not to come home empty-handed ". João Fogaça, the Governor of S.Jorge da Mina, placed on board his vessel the gold he had obtained by barter.

Ultimately, in December, 1488, after an absence of sixteen months and seventeen days, he once more entered the River Tagus, on his way to Lisbon. Bartolomeu Dias had discovered 373 legoas or 1260 miles of coast.

His voyage, jointly with the reports received by that time from Pero do Covilhã, had demonstrated the fact that India might be reached by sea.

We are not aware that Bartolomeu Dias ever received a reward for his great achievement. It seems not, for between 1490 and 1495 he still commanded the Christovão, and when King João had overruled the objections of his advisers, who thought it unwise to expend, and possibly exhaust, the resources of the kingdom in distant adventures, which, even if successful, would raise against Portugal all those who now profited, or who in the future hoped to profit, from the India trade, it was not Bartolomeu Dias who was placed at the head of the expedition that was to crown the enterprise started by Prince Henry.

CHAPTER 5

THE FIRST VOYAGE OF COLUMBUS (1492-1493)

On his return, Bartolomeu Dias could have been considered the world's greatest discoverer. However, that title was soon to be taken away from him by Christopher Columbus, who was present when Bartolomeu Dias described his voyage of discovery to the King of Portugal.

Christopher Columbus took a particular interest in Bartolomeu Dias' discovery of the sea route to the east. Columbus owned one of the few remaining references to the voyage, which is in his own handwriting.

Despite his voyages of discovery, Bartolomeu Dias took no part in the voyages of Christopher Columbus. Bartolomeu Dias was to wait his turn before facilitating further discoveries.

Whilst Bartolomeu Dias had been pioneering an eastern sea route to Asia, Christopher Columbus had been trying to arrange a voyage to Asia via a western sea route.

Although Christopher Columbus had been invited by the King of Portugal to Lisbon in 1488, in time to see the return of Bartolomeu Dias, his discussions with the Portuguese government did not facilitate his voyage to Asia via a western sea route. So Christopher Columbus returned to Spain.

Unable throughout 1490 to get a hearing at the Spanish court, Christopher Columbus was in 1491 again referred to a junta, presided over by Cardinal Mendoza. However, this junta, to Columbus' dismay, once more rejected his proposals. The Spanish sovereigns merely promised him that when the Granada war was over, they would reconsider what he had laid before them.

Columbus was now in despair. He at once betook himself to Huelva, a little maritime town in Andalusia, north-west of Cadiz, with the intention of taking ship for France. He halted, however, at the monastery of La Rabida, near Huelva, and still nearer Palos, where he seems to have made lasting friendships on his first arrival in Spain in January 1485. Whilst there he enlisted the support of Juan Perez, the guardian, who invited him to take up his quarters in the monastery, and introduced him to Garcia Fernandez, a physician and student of geography. Juan Perez had been the queen's confessor. He now wrote to her in urgent terms, and was summoned to her presence. Money was sent to Columbus to bring him once more to court. He reached Granada in time to witness the surrender of the city on 2nd January 1492, and negotiations were resumed.

Columbus believed in his mission, and stood out for high terms. He asked for the rank of admiral at once ("Admiral of the Ocean" in all those islands, seas, and continents that he might discover), the vice-royalty of all he should discover, and a tenth of the precious metals discovered within his admiralty. These conditions were rejected, and the negotiations were again interrupted. An interview with Mendoza appears to have followed, but nothing came of it, and before the close of January 1492, Columbus actually set out for France.

At length, however, on the entreaty of the Queen's confidante, the Marquesa de Moya, of Luis de Santangel, receiver of the ecclesiastical revenues of the crown of Aragon, and of other courtiers, Isabella was induced to determine on the expedition. A messenger was sent after Columbus, and overtook him near a bridge called "Pinos" near Granada. He returned to the camp at Santa Fé; and on the 17th of April 1492, the agreement between him and their Catholic majesties was signed and sealed.

As his aims included not only the discovery of Cipangu or Japan, but also the opening up of intercourse with the grand khan of Cathay, Columbus received a royal letter of introduction to the latter. The town of Palos was ordered to find him two ships, and these were soon placed at his disposal. However, no crews could be got together, in spite of the indemnity offered to criminals and "broken men" who would serve on the expedition. Had not Juan Perez succeeded in interesting in the cause the Palos "magnates" Martin Alonso Pinzon and Vicente Yañez Pinzon, Columbus' departure had been long delayed. At last, however, men, ships and stores were ready. The expedition consisted of the

"Santa Maria," a decked ship of 100 tons with a crew of 52 men, commanded by the admiral in person; and of two caravels; the "Pinta" of 50 tons, with 18 men, under Martin Pinzon; and the "Niña," of 40 tons, with 18 men, under his brother Vicente Yañez, afterwards (1499) the first to cross the line in the American Atlantic.

The adventurers numbered 88 souls. On Friday, the 3rd August 1492, at eight in the morning, the little fleet weighed anchor, and stood for the Canary Islands. An abstract of the admiral's diary made by Las Casas is yet extant. From it many particulars may be gleaned concerning this first voyage. Three days after the ships had set sail the "Pinta" lost her rudder. The admiral was in some alarm, but comforted himself with the reflection that Martin Pinzon was energetic and ready witted. They had, however, to put in at Teneriffe, to refit the caravel.

On the 6th of September they weighed anchor once more with all haste, Columbus having been informed that three Portuguese caravels were on the look-out to intercept him.

On 13th of September the westerly variations of the magnetic needle were for the first time observed.

On the 15th a meteor fell into the sea at four or five leagues distance. Soon after they arrived at those vast plains of seaweed called the Sargasso Sea. All the time, writes the admiral, they had most temperate breezes, the sweetness of the mornings being especially delightful, the weather like an Andalusian April, and only the song of the nightingale wanting.

On the 17th September the men began to murmur. They were frightened by the strange phenomena of the variation of the compass, but the explanation Columbus gave restored their tranquillity.

On the 18th they saw many birds, and a great ridge of low-lying cloud, and so they expected to see land.

On the 20th they saw boobies and other birds, and were sure the land must be near. In this, however, they were disappointed, and thenceforth

Columbus, who was keeping all the while a double reckoning, one for the crew and one for himself, had great difficulty in restraining the evil-disposed from the excesses they meditated.

On the 25th September, Martin Alonso Pinzon raised the cry of land, but it proved false, as did the rumour to the same effect on the 7th October, from the "Niña."

However, on the 11th October the "Pinta" fished up a cane, a pole, a stick which appeared to have been wrought with iron, and a board, while the "Niña" sighted a branch covered with berries; "and with these signs all of them breathed and were glad." At ten o'clock on that night Columbus himself perceived and pointed out a light ahead, and at two in the morning of Friday, the 12th of October 1492, Rodrigo de Triana, a sailor aboard the "Niña," announced the appearance of what proved to be the New World. The land sighted was an island, called by the Indians Guanahani, and named by Columbus San Salvador. It is generally identified with Watling Island.

The same morning Columbus landed, richly clad, and bearing the royal banner of Spain. He was accompanied by the brothers Pinzon, bearing banners of the Green Cross (a device of the admiral's), and by great part of the crew. When they all had "given thanks to God, kneeling upon the shore, and kissed the ground with tears of joy, for the great mercy received," the admiral named the island, and took solemn possession of it for their Catholic majesties of Castile and Leon. At the same time such of the crews as had shown themselves doubtful and mutinous sought his pardon weeping, and prostrated themselves at his feet.

It resulted in the discovery of the islands of Santa Maria de la Concepcion (Rum Cay), Fernandina (Long Island), Isabella (Crooked Island), Cuba or Juana (named by Columbus in honour of the young prince of Spain), and Hispaniola, Haiti, or San Domingo. Off the last of these the "Santa Maria" went aground, owing to the carelessness of the steersman. No lives were lost, but the ship had to be unloaded and abandoned. Columbus, who was anxious to return to Europe with the news of his achievement, resolved to plant a colony on the island, to build a fort out of the material of the stranded hulk, and to leave the crew. The fort was called La Navidad. 44 Europeans were placed in charge of it. On the 4th January 1493 Columbus, who had lost sight of Martin Pinzon, set sail alone in the "Niña" for the east.

Two days later the "Pinta" joined her sister-ship. A storm, however, separated the vessels, and it was not until the 18th February that Columbus reached the island of Santa Maria in the Azores. Here he was threatened with capture by the Portuguese governor, who could not for some time be brought to recognise his commission. On the 24th February, however, he was allowed to proceed, and on the 4th March the "Niña" dropped anchor off Lisbon. The king of Portugal received the admiral with the highest honours. On the 13th of March the "Niña" put out from the Tagus, and two days afterwards, Friday, the 15th of March, she reached Palos.

The court was at Barcelona. After despatching a letter announcing his arrival, Columbus proceeded in person, and when he entered the city in a sort of triumphal procession, was received by their majesties in full court, and, seated in their presence, related the story of his wanderings, exhibiting the "rich and strange" spoils of the new-found lands : the gold, the cotton, the parrots, the curious arms, the mysterious plants, the unknown birds and beasts, and the Indians he had brought with him for baptism.

All his honours and privileges were confirmed to him : the title of Don was conferred on himself and his brothers, he rode at the king's bridle, he was served and saluted as a grandee of Spain. A new and magnificent scutcheon was also blazoned for him (4th May 1493), whereon the royal castle and lion of Castile and Leon were combined with the five anchors of his own coat of arms. Nor were their Catholic highnesses less busy on their own account than on that of their servant. On the 3rd and 4th of May Alexander VI granted bulls confirming to the crowns of Castile and Leon all the lands discovered, to be discovered, west of a line of demarcation drawn 100 leagues west of the Azores, on the same terms as those on which the Portuguese held their colonies along the African coast. A new expedition was got in readiness with all possible despatch, secure and extend the discoveries already made.

CHAPTER 6

THE SECOND VOYAGE OF COLUMBUS (1493-1496)

After several delays the fleet weighed anchor on the 24th September 1493 and steered westwards. It consisted of three great carracks (galleons) and fourteen caravels (light frigates), having on board over 1500 men, besides the animals and materials necessary for colonisation.

Twelve missionaries accompanied the expedition, under the orders of Bernardo Buil or Boil, a Benedictine. Columbus had already been directed on 29th May 1493 to endeavour by all means in his power to Christianise the inhabitants of the islands, to make them presents, and to "honour them much". All under him were commanded to treat them "well and lovingly," under pain of severe punishment.

On 13th October the ships, which had put in at the Canaries, left Ferro, and on Sunday 3rd November, after a single storm, "by the goodness of God and the wise management of the admiral" an island was sighted to the west that was named Dominica. Northwards from this the isles of Marigalante and Guadalupe were next discovered and named, while on the north-western course to La Navidad those of Montserrat, Antigua, San Martin, Santa Cruz and the Virgin Islands were sighted, and the island now called Puerto Rico was touched at, hurriedly explored, and named San Juan Bautista.

On the 22nd November Columbus came in sight of Hispaniola, and sailing westward to La Navidad, found the fort burned and the colony dispersed. He decided on building a second fort. Coasting onwards east of Monte Cristi, he pitched on a spot where he founded the city of Isabella.

The climate proved unhealthy. The colonists were greedy of gold, impatient of control, proud, ignorant and mutinous. Columbus, whose inclination drew him westward, was doubtless glad to escape the worry and anxiety of his post, and to avail himself of the instructions of his sovereigns as to further discoveries. On the 2nd of February 1494 he sent home, by Antonio de Torres, that despatch to their Catholic highnesses by which he may be said to have founded the West Indian slave trade.

He established the mining camp of San Tomaso in the gold country of Central Hispaniola. On the 24th April 1494, having nominated a council of regency under his brother Diego, and appointed Pedro Margarit his captain-general, he again put to sea. After following the southern shore of Cuba for some days, he steered southwards, and on 14th May discovered the island of Jamaica, which he named Santiago.

Columbus then resumed his exploration of the Cuban coast, threaded his way through a labyrinth of islets that he named the Garden of the Queen (Jardin de la Reyna), and, after coasting westwards for many days, became convinced that he had discovered continental land. He therefore caused Perez de Luna, the notary, to draw up a document to this effect (12th of June 1494), which was afterwards taken round and signed (the admiral's steward witnessing) by the officers, men and boys of his three caravels, the "Niña," the "Cordera," and the "San Juan."

He then stood to the south-east, and sighted the island of Evangelista (now Isla de los Pinos), revisited Jamaica, coasted the south of Hispaniola, and on the 24th of September touched at and named the island of La Mona, in the channel between Hispaniola and Puerto Rico.

Thence he had intended to sail eastwards and complete the survey of the Caribbean Archipelago; but he was exhausted by the terrible wear and tear of mind and body he had undergone. He says himself that on this expedition he was three-and-thirty days almost without sleep. On the day following his departure from La Mona he fell into a lethargy, that deprived him of sense and memory, and had well-nigh proved fatal to life.

At last, on the 29th of September, the little fleet dropped anchor off Isabella, and in his new city the admiral lay sick for five months.

The colony was in a sad plight. Everyone was discontented and many were sick, for the climate was unhealthy and there was nothing to eat.

Margarit and Boil had deserted the settlement and had fled to Spain. However, before his departure, the former, in his capacity of captain-general, had done a lot to outrage and alienate the Indians. The strongest measures were necessary in order to undo this mischief. Backed by his brother Bartholomew, Christopher Columbus proceeded to reduce the natives under Spanish sway. Alonso de Ojeda succeeded by a brilliant coup de main in capturing the cacique Caonabo, and the rest submitted.

Five ship-loads of Indians were sent off to Seville on 24th June 1495 to be sold as slaves. A tribute was imposed upon their fellows, which must be looked upon as the origin of that system of repartimientos or encomiendas that was afterwards to work such mischief among the conquered.

In October 1495 Juan Aguado arrived at Isabella, with a royal commission to report on the state of the colony. Here he took up the position of a judge of Columbus's government, and much recrimination followed. Columbus decided to return home. He appointed his brother Bartholomew adelantado of the island, and on 10th March 1496 he quit Hispaniola in the "Niña."

The vessel, after a protracted and perilous voyage, reached Cadiz on 11th June 1496, where the admiral landed, wearing the habit of a Franciscan. He was cordially received by his sovereigns, and a new fleet of eight vessels was put at his disposal.

By royal patent, moreover, a tract of land in Hispaniola, of 50 leagues by 20, was offered to him, with the title of duke or marquis, which he declined. For three years he was to receive an eighth of the gross and a tenth of the net profits on each voyage; the right of creating a mayorazgo or perpetual entail of titles and estates was granted him; and his two sons were received into Isabella's service as pages.

The voyages of Christopher Columbus to the west on behalf of Spain had completely ousted any consideration of Bartolomeu Dias still being considered the world's greatest discoverer. However, even before Christopher Columbus had returned to Spain, Bartolomeu Dias was already preparing further voyages of discovery to the east on behalf of Spain's great rival Portugal.

CHAPTER 7

THE VOYAGE OF VASCO DA GAMA

Bartolomeu Dias had found the sea-gates of the Orient. It remained for some mariner of equal daring to force them open. Wars with Castile and the death of King João II had delayed this venture for a decade, but King Manoel, who succeeded to the throne in 1495, did not long hesitate to resume the historic mission bequeathed to his country by Prince Henry the Navigator. This had now come to mean the search for a sea route to India.

The twofold purpose of the quest was explained with admirable brevity by the first Portuguese sailor who disembarked on Indian soil. "Christians and spices", he replied, when asked what had brought him and his comrades so far.

All those who still cherished the crusading ideals of a bygone age dreamed of an alliance with Prester John's empire and with the other Catholic powers that were believed to exist on the other side of the world. This accomplished, the chivalry of Portugal would lead the united hosts of European and Asiatic Christendom in a campaign for the destruction of Muhammadanism. Others hoped to divert for their own profit the trade in Indian wares, and especially in spices, which had hitherto filled the treasuries of Genoa, Venice and Ragusa.

Shortly after his accession, King Manoel summoned to his court at Estremoz the son of a certain Estevão da Gama, who had been chosen to lead the way to India but had died while the preparations for the voyage were still incomplete. His third son Vasco was appointed in his stead to the office of Captain-Major (Capitão-Mór) or Commander-in -chief. Castanheda states that the honour was first offered to Vasco's eldest brother, Paulo da Gama, who declined it on the ground of ill-health.

When Vasco da Gama was chosen for the Indian voyage, he was already an expert navigator, and was unmarried. Courage, ambition, pride and unwavering steadfastness of purpose were the bedrock of his character. Although on occasion he might unbend so far as to join his sailors in a hornpipe, he allowed no relaxation of discipline. Although he made promotion depend exclusively on merit, never on the fortune of birth - "preferring", as Corrêa puts it, "a low man who had won honour with his right arm to a gentleman Jew" - he was at heart an aristocrat.

Early in the summer of 1497 he was granted an audience of King Manoel at Montemór-o-Novo, near Evora, where he took the oath of fealty, and was presented with a silken banner emblazoned with the Cross of the Order of Christ. He then journeyed to Lisbon to assume command of four ships that already lay moored in the Tagus estuary.

Two sister ships, the São Rafael and São Gabriel, had been built expressly for this voyage. Bartolomeu Dias was their architect. Dias had set himself to design a vessel better adapted than the caravel type for a long cruise in stormy latitudes. Bartolomeu Dias was compelled to sacrifice some good qualities of the older vessel - its speed, its handiness in working to windward, its finer lines. However, the new ships were strong and seaworthy enough to hold their own among the greybeards of the South Atlantic, and roomy enough to accommodate men and officers without overmuch discomfort. Low amidships, with high castles towering fore and aft, they rode the water like ducks - square-sterned, bluff-bowed, their length about thrice their beam. Each had three masts, the fore and main carrying two square sails apiece, while the mizzen bore a single lateen sail. The bowsprit was tilted upwards at so high an angle that it resembled a fourth mast, fitted with one square-sail.

Vasco da Gama had chosen the São Gabriel as his flagship, while Paulo da Gama commanded the São Rafael. The flotilla was completed by the Berrio, commanded by Nicolau Coelho, and a store ship under a retainer of Vasco da Gama named Gonçalo Nunes.

Vasco da Gama, Bartolomeu Dias and their advisers had done their utmost to organise success by giving a technical training to the crews, providing stores for three years, and securing the best scientific outfit available. According to Corrêa, "Vasco da Gama spoke to the sailors who were told off for the voyage, and strongly recommended them, until the time of their departure, to endeavour to learn to be carpenters, rope-makers, caulkers, blacksmiths, and plank-makers; and for this purpose he gave them an increase of two cruzados a month beyond the sailors' pay that they had, which was of five cruzados a month, so that all rejoiced at learning, so as to draw more pay. And Vasco da Gama bought for them all the tools that befitted their crafts".

Tables showing the declination of the sun were provided by the astronomer-royal, Abraham Zacuto ben Samuel. These, which enabled navigators to determine their latitude by calculating the altitude of the sun when the pole-star was invisible, had been translated from Hebrew into Latin in the previous year, and printed at Leiria under the title of Almanach Perpetuum Celestium motuum cujus radix est 1473.

Other books, maps, and charts were supplied by D. Diogo Ortiz de Vilhegas, titular Bishop of Tangier and (as Bishop of Ceuta) one of the three royal commissioners who had discredited Columbus's plans for a voyage to Cipangu under the Portuguese flag. Among these documents were, almost certainly, the Geography of Ptolemy, the Book of Marco Polo, and copies of the reports sent home by the Jew Pedro de Covilhã and other Portuguese explorers who had been sent overland to Asia, besides a transcript of the information furnished by Lucas Marcos, an Abyssinian priest who visited Lisbon in 1490. The log and charts of Bartolomeu Dias were of course available, conceivably also the map of Henricus Martellus Germanus.

CHAPTER 8

THE FLEET OF VASCO DA GAMA

All authorities agree that the fleet, or armada, fitted out for the voyage numbered four vessels, but they are not agreed as to the names that these vessels bore. We are not, however, likely to be misled if we accept the unanimous testimony of the author of our Roteiro, of João de Barros, Lopez de Castanheda, Pedro Barretto de Rezende, and Manuel Faria y Sousa, according to whom the names of the ships and of their principal officers were as follows : -

São Gabriel (the flagship) - Vasco da Gama was the captain-major. Pero d'Alenquer was the pilot. Gonçalo Alvarez was the master. Diogo Dias was the clerk.

São Rafael - Paulo da Gama was the captain. João de Coimbra was the pilot. João de Sá was the clerk.

Bérrio - Nicolau Coelho was the captain. Pero Escolar was the pilot. Alvaro de Braga was the clerk.

Store-ship - Gonçalo Nunes was the captain.

Correa and the unknown author of the Jornal das Viagens call the "Berrio" São Miguel, and make the São Rafael the flagship. L. de Figueiredo de Falcão substitutes a São Miguel for the São Rafael. It is just possible that the vessel popularly called Bérrio, after its former owner, had been re-christened São Miguel.

The Bérrio was one of those swift lateen-rigged vessels for which Portugal was famous from the thirteenth century. Their burthen did not exceed 200 tons, and they had two or three masts, and occasionally even four. The Bérrio is stated to have been a vessel of only 50 tons. She was named after her former owner and pilot, of whom she was purchased expressly for this voyage. The store-ship was of more considerable size. Sernigi says she measured 110 tons. Castanheda credits her with 200 tons. She may have been a so-called caravela redonda, that is a caravel that carried square sails on the main and foremasts and triangular ones on the mizzenmast and the bowsprit. This vessel was purchased of Ayres Correa, a ship owner of Lisbon.

The São Gabriel and São Rafael were specially built for this voyage. Bartolomeu Dias, who superintended their construction, discarded the caravel in which he had achieved his great success, in favour of square-rigged vessels of greater burthen, which, although slower sailers and less able to ply to windward, offered greater safety and more comfort to their crews. Bartolomeu Dias took care, at the same time, that the draught of these vessels should enable them to navigate shallow waters, such as it was expected would be met with in the course of the voyage. The timber for these two vessels had been cut during the last year of the reign of King João, in the Crown woods of Leiria and Alcacer. The vessels having been completed, the King ordered them to be equipped by Fernão Lourenço, the factor of the house of Mines, and one of the most "magnificent" men of his time.

No contemporary description or picture of these vessels has reached us, but there can hardly be a doubt that their type is fairly represented on a painting made by order of D.Jorge Cabral, who was Governor of India from 1549 to 1550. This painting subsequently became the property of D. João de Castro. A copy of it was first published by the Visconde de Juromenha, who took it from a MS. dated 1558. The fine woodcut in W.S.Lindsay's History of Merchant Shipping (II, page 5), from an ancient picture that also belonged to D.João de Castro, seems to be derived from the same source, but as the vessel carries the flag of the Order of Christ at the main, and not the Royal Standard, it cannot represent the flagship. At all events, it is not more authentic than either of the ships delineated in the drawing first published by Juromenha.

Authorities differ very widely as to the tonnage of these two vessels. Sernigi says they were of 90 tons each, thus partly bearing out Correa, who states that the three ships (including the Bérrio) were built of the same size and pattern. D.Pacheco Pereira states that the largest of them did not exceed 100 tons. J.de Barros gives them a burthen of between 100 and 120 tons, whilst Castanheda allocates 120 tons to the flagship and 100 to the São Rafael.

Whilst the authorities quoted dwell upon the small size of the vessels, which for the first time reached India from a European port, and even give reasons for this limitation of burthen, there is some ground for believing that the tonnage of the ships, expressed according to modern terminology, was in reality much greater than is usually supposed. Pedro Barretto de Rezende may therefore have some justification when he states that these vessels ranged from 100 to 320 tons. Mr. Lindsay would go even further. The São Gabriel, according to him, was constructed to carry 400 pipes, equivalent to 400 tons measurement, or about 250 to 300 tons register. He adds that E. Pinto Bastos agrees with him.

In considering this question of tonnage, it must be borne in mind that "ton", at the close of the fifteenth century, was a different measure from what it is at present. We learn from E.A.D'Albertis that the tonelada of Seville was supposed to afford accommodation for two pipes of 27½ arrobas (98 gallons) each, and measured 1.405 cubic metres, or about 50 cubic feet. The tonel of Biscay was 20 per cent larger. According to Captain H.Lopez de Mendonça, the tonel at Lisbon measured 6 palmos de goa in length (talha), and 4 such palmos in breadth and height (parea), that is, about 85 cubic feet. This, however, seems excessive, because two butts of sherry of 108 gallons each would occupy only 75 cubic feet. At any rate, these data show that the ton of the fifteenth century was considerably larger than the ton measurement at present.

Two attempts have been made by officers of the Portuguese Navy, Captains João Braz d'Oliveira and A.A.Baldaque da Silva, to reconstruct the flagship, or rather to design a ship of a type existing at the close of the fifteenth century, and answering as nearly as possible to the scanty indications to be found within the pages of the historians of this memorable voyage. In this reconstruction good use was made of an early manuscript on shipbuilding by Fernando Oliveira (O livro da fabrica das Nâos).

The designs produced by the two naval officers differ widely in several respects, and more especially as regards the relation between the total length of the ship and the breadth of beam. In Captain B. da Silva's ship, the beam is equal to one third of the length, whilst the proportion in Captain J.Braz d'Oliveira's ship is as one to five. The former of these ships is broad-beamed, as befits the period, whilst the latter is almost as slim as a modern clipper. It must be remembered that it was held that the length of a sailing ship should not exceed four times the breadth of beam, and this maxim was undoubtedly acted upon by the shipbuilders of the fifteenth century.

Captain Baldaque da Silva's design of the São Gabriel has been embodied in a model. The dimensions of the ship designed by Captain Baldaque da Silva are (in feet) as follows :- Length over all - 84.1, Water-line (when laden) - 64.0, Keel - 56.7, Breadth of beam - 27.9, Depth - 17.1, Draught (abaft) - 7.5, Draft (forward) - 5.6, Metocentric height above the water-line (laden) - 7.4, Displacement - 178 tons, Tonnage - 4130 cubic feet or 103 tons.

This, as I learn from a private letter of Captain B. da Silva, is supposed to be the gross under-deck tonnage, but on calculating the tonnage according to the Builders Old Measurement Rule, I find it to amount to 230 tons of 40 cubic feet each, whilst the "expeditive" method practised at Venice during the fifteenth century yields 896 botte of 28 gallons, or about 250 toneladas.

The ship was flat-bottomed, with a square stern and bluff bow, the latter ornamented with a figure of her patron saint. Wales were placed along the sides to reduce her rolling when going before the wind. Formidable "castles" rose fore and aft, having a deep waist between them. These "castles", however, had not then grown to the portentous height attained at a subsequent period, when they rendered it difficult to govern the ship in a gale, and it often became necessary to cut down the foremast and dismantle the forecastle to enable them to keep her head to the wind.

These "castles" were in reality citadels, and enabled the crew to make a last stand after the vessel had been boarded. A notable instance of this occurred in the course of the fight with the Meri in 1502.

The captain was lodged in the castle rising upon the quarter-deck. The officers were accommodated in the room below his and in the forecastle, whilst the men had their quarters beneath the gang boards, which ran along the topsides from castle to castle. The men were each allowed a locker, to contain such goods as they might obtain by barter with the natives. Ladders led from the main deck up to the fighting decks of the two castles, and these were defended against boarders by nettings. The tiller of the rudder entered the battery abaft the captain's apartments, where also stood the binnacle. The armament consisted of twenty guns. The lower battery of the "castle" rising on the quarter-deck was armed with eight breech-loaders made of wrought -iron staves, held together by hoops and mounted on forked props. The upper battery held six bombards, and the forecastle the same number. We may at once state that the men carried no firearms. Their arms included crossbows, spears, axes, swords, javelins, and boarding-pikes. Some of the officers were clad in steel armour, whilst the men had to be content with leather jerkins and breastplates.

Amidships stood the batel, or long boat, in addition to which there was available a yawl rowed with four or six oars. There were three masts and a bowsprit. The mainmast rose to a height of 110 feet above the keel and flew the Royal Standard at its head, whilst the captain's scarlet flag floated from the crow's-nest, nearly 70 feet above deck. A similar crow's-nest was attached to the foremast. In the case of an engagement these points of vantage were occupied by fighting men, who hurled thence javelins, grenades, and powder-pots upon the enemy.

The sails were square, with the exception of that of the mizzen, which was triangular. When spread they presented 4,000 square feet of canvas to the wind. This was exclusive of the "bonnets" that were occasionally laced to the leeches of the mainsail, and served to some extent the same purpose as a modern studding-sail. The Cross of the Order of Christ was painted on each sail.

The anchors, two in number, were of iron, with a wooden stock and a ring for bending the cable. The hold was divided into three compartments. Amidships were the water barrels, with coils of cable on the top of them - a very inconvenient arrangement. Abaft was the powder-magazine, and most arms and munitions, including iron and stone balls, were kept there. The forward compartment was used for the storage of requisites, including spare sails and a spare anchor.

The lower deck was divided by bulkheads into three compartments, two of which were set apart for provisions, presents, and articles of barter. The "provisions", according to Castanheda, were calculated to suffice for three years, and the daily rations were on a liberal scale, consisting of 1½ pounds of biscuit, 1 pound of beef or half a pound of pork, 2½ pints of water, 1¼ pints of wine, one-third of a gill of vinegar, and half that quantity of oil. On fast days, half a pound of rice, of codfish, or cheese was substituted for the meat. There were, in addition, flour, lentils, sardines, plums, almonds, onions, garlic, mustard, salt, sugar and honey. These ship's stores were supplemented by fish, caught whenever an opportunity offered, and by fresh provisions obtained when in port, among which were oranges, which proved most acceptable to the many men suffering from scurvy.

The merchandise was not only insufficient in quantity, but proved altogether unsuited to the Indian market. It seems to have included lambel (striped cotton stuff), sugar, olive-oil, honey, and coral beads. Among the objects intended for presents, there were wash-hand basins, scarlet hoods, silk jackets, pantaloons, hats, Moorish caps; besides such trifles as glass beads, little round bells, tin rings and bracelets, which were well enough suited for barter on the Guinea coast, but were not appreciated by the wealthy merchants of Calicut. Of ready money there seems to have been little to spare. All this is made evident by the letters of Dom Manuel and Signor Sernigi.

The scientific outfit of the expedition, it may safely be presumed, was the best to be procured at the time. The learned D.Diogo Ortiz de Vilhegast furnished Da Gama with maps and books, including, almost as a matter of course, a copy of Ptolemy, and copies of the information on the East collected at Lisbon for years past. Among these reports, that sent home by Pero de Covilhão found, no doubt, a place, as also the information furnished by Lucas Marcus, an Abyssinian priest who visited Lisbon about 1490.

The astronomical instruments were provided by Zacut, the astronomer, and it is even stated that Vasco enjoyed the advantage of being trained as a practical observer by that learned Hebrew. These instruments included a large wooden astrolabe, smaller astrolabes of metal, and, in all probability, also quadrants. They were accompanied by a copy of Zacut's Almanach perpetuum Celestium motuum cujus radix est 1473, a translation of which, by José Vizinho, had been printed at Leiria in 1496. These tables enabled the navigator to calculate his latitudes by observing the altitude of the sun.

There was, of course, a sufficient supply of compasses, of sounding leads and hour-glasses, and possibly also a catena a poppa, that is, a rope towed at the stern to determine the ship's leeway, and a toleta de marteloia, a graphical substitute for traverse tables, both of them contrivances long since in use among the Italians. It is also possible that Vasco da Gama was already provided with an equinoctial compass for determining the time of high water at the ports he visited, and with a variation compass. This instrument consisted of a combination of a sundial with a magnetic needle. It had been invented by Peurbach, in around 1460. It was improved by Felipe Guillen in 1528, and by Pedro Nunes in 1537, and was used for the first time on an extensive scale by João de Castro, during a voyage to India and the Red Sea, in 1538-1541. We are inclined to think that Vasco had such a variation compass, for the Cabo das Agulhas, or "Needle Cape", thus named because the needle there pointed, or was supposed to point, due north, has already found a place on Cantino's Chart, and can have been named only as the result of an actual observation, however inaccurate.

Lastly, there remain to be noticed the Padrãos, or pillars of stone that were on board the vessels, and three of which, by the king's express desire, were dedicated to São Rafael, São Gabriel, and Santa Maria. Barros and Castanheda tell us that these pillars resembled those set up by Diogo Cão and Bartolomeu Dias in the time of Dom João II, and in a series of pictures that Dom Manuel desired to have painted in celebration of the discovery of India, the Padrão to be shown at the Cape of Good Hope, or "Prasum promontorium", was to have been surmounted by a cross, and to bear the Royal Arms and a Pelican, with an inscription giving the date.

Correa, on the other hand, affirms in his Lendas that the pillar set up at the Rio da Misericordia (the Rio dos Bons Signaes of the Roteiro) was of marble, with two escutcheons, one of the arms of Portugal, and the other (at the back) of a sphere, and that the inscription was “Do senhorio de Portugal reino de Christãos”. The pillar at Melinde had the same escutcheons, but the inscription was limited to the words “Rey Manoel”. As Correa had an opportunity of seeing these pillars, his description of them may be correct, though he is an arrant fabulator.

CHAPTER 9

THE CREW OF VASCO DA GAMA

The officers and men in the armada were carefully selected. Several of them had been with Bartolomeu Dias around the Cape. All of them justified by their conduct, under sometimes trying circumstances, the selection that had been made.

Authorities widely differ as to the number of men who embarked. Sernigi says there were only 118, of whom 55 died during the voyage and only 63 returned. Galvão says there were 120, besides the men in the store ship. Castanheda and Goes raise the number to 148, of whom only 55 returned, many of them broken in health. Faria y Sousa and San Ramon say there were 160, and the latter adds that 93 of these died during the voyage, thus confirming a statement made by King Manuel in his letter of 20th February 1504, to the effect that less than one-half returned. According to Barros there were 170 men, including soldiers and sailors. Correa raises the number to 260, for he says that in each of the three ships there were 80 officers and men, including servants, besides six convicts and two priests. He says nothing of the store-ship. By the time Vasco da Gama had reached the Rio da Misericordia only 150 out of this number are said to have been alive.

Correa, no doubt, exaggerates. On the other hand, Sernigi's numbers seem to us to err quite as much on the other side. It is quite true that a Mediterranean merchantman of 100 tons, in the sixteenth century, was manned by 12 able and 8 ordinary seamen, but in the case of an expedition sent forth for a number of years and to unknown dangers, this number would no doubt have been increased. We are, therefore, inclined to believe that the number given by De Barros - namely, 170 - may be nearer the truth, namely 70 men in the flagship, 50 in the São Rafael, 30 in the caravel, and 20 in the store-ship. The men in the flagship may have included 1 captain, 1 master, 1 pilot, 1 assistant pilot, 1 mate (contramestre), 1 boatswain (guardião), 20 able seamen (marinheiros), 10 ordinary seamen (grumetes), 2 boys (pagens), 1 chief gunner or constable, 8 bombardiers, 4 trumpeters, 1 clerk or purser (escrivão), 1 storekeeper (dispenseiro), 1 officer of justice (meirinho), 1 barber-surgeon, 2 interpreters, 1 chaplain, 6 artificers (ropemaker, carpenter, calker, cooper, armourer and cook), and 10 servants. One or more of these servants may have been African slaves. The “Degradados”, or convicts on board, to be “adventured on land”, are included in the total. Whether private gentlemen were permitted to join this expedition as volunteers, history does not record.

The following “muster-roll” contains short notices of all those who are stated to have embarked at Lisbon in Vasco da Gama’s fleet, or who subsequently joined it, either voluntarily or upon compulsion.

Apart from natives, 31 people are mentioned. With respect to 26 of these no reasonable doubt can be entertained that they were actually members of the ships’ companies.

Those among them whose names appear in the “Journal” are distinguished by an asterisk.

Captains :

Vasco da Gama, Captain-Major in the São Gabriel.

Paulo da Gama, his brother, commanding the São Rafael.

Nicolau Coelho, Captain of the Berrio or São Miguel. He subsequently went out to India with Cabral in 1500, and for a third time with Francisco d’Albuquerque, in 1503. On 24th February 1500, the King granted him a pension of 70,000 reis. He also received a coat-of-arms : a field gules, charged with a lion rampant between two pillars (padrãos), silver, standing upon hillocks by the sea vert, and two small escutcheons charged with five bezants. He seems to have been dead in 1522, for on 19th December 1522, his son, Francisco, begged the King to transfer the pension of his late father to himself.

Gonçalo Nunes, Captain of the store-ship. He was a retainer of Vasco da Gama.

Pilots and masters :

Pero d'Alenquer, pilot of the São Gabriel. He had been with Bartolomeu Dias in the discovery of the Cape of Good Hope, and with the Congo mission in 1490.

João de Coimbra, pilot of the São Rafael. An African slave belonging to him deserted at Mozambique.

Pero Escolar, pilot of the Berrio. On February 18th 1500, the King granted him a pension of 4,000 reis. He went as pilot with Pedro Cabral.

Gonçalo Alvares, master of the São Gabriel. He subsequently held the office of pilot-major of India. On January 26th 1504, the King granted him a pension of 6,000 reis.

André Gonçalves. According to Correa, he had been with Vasco da Gama, whose interest had procured him an appointment in Pedro Cabral's fleet. The same untrustworthy author states that Cabral sent him back from Brazil with the news of his discovery, and that the King immediately after his arrival, fitted out a fleet to continue the explorations in the New World. Barros and Castanheda state that Cabral sent back Gaspar de Lemos. Neither they, nor, as far as I am aware, any other authority, mention an André Gonçalves in connection with Gama's or Cabral's expeditions.

Purser or clerks :

Diogo Dias, clerk of the São Gabriel. He was a brother of Bartolomeu Dias, the discoverer of the Cape of Good Hope.

João de Sá, clerk of the São Rafael. He also went to India with Cabral, and subsequently became treasurer of the India House.

Alvaro de Braga, clerk of the Berrio. Vasco appointed him head of the factory at Calecut. Correa erroneously calls him Pedro de Braga. He was rewarded by the King on February 1st 1501.

Interpreters :

Martim Affonso. He had lived in Congo.

Fernão Martins. Vasco sent him to the King of Calecut, and he was present at the audience that Vasco da Gama had of the King. Subsequently he filled several positions of trust in India. He is the “African slave” who spoke Arabic, referred to by Correa.

João Martins.

Priests :

Pedro de Covilhã, called Pero de Cobillones by Faria y Sousa, who refers to ancient documents and the assertion of F. Christoval Osorio, of the Order of Trinity, as his authorities. He was Prior of a monastery of the Order of the Trinity at Lisbon, and went out as Chaplain of the Fleet and Father Confessor. According to Francisco de Sousa’s Oriente Conquistado, he died a martyr on July 7th 1498, and this statement is accepted by P. Francisco de S. Maria. Fr. Jeronymo de São Jose enlarges upon by stating that this apocryphal “protomartyr” of India “was speared whilst expounding the doctrines of the Trinity”. At the date of his alleged death, Vasco da Gama was still at Calecut. He may have died of disease. Neither Barros, Castanheda, nor Correa mentions the name of this priest.

João Figueiro. Correa claims to have derived much information from a diary kept by this priest, of which only fragments appear to have come into his possession. Other authors ignore the name.

Sailors and soldiers :

João d’Ameixoeira or Dameiroeiro. According to Correa, he was one of the mutineers who returned to Portugal. No other writer mentions him.

Pedro de Faria e Figueiredo, died at Cabo das Correntes.

Francisco de Faria e Figueiredo, brother of the preceding. He wrote Latin verses. He, too, died at Cabo das Correntes.

Sancho Mexia, incidentally mentioned in the Roteiro.

João Palha, one of the thirteen who attended Vasco da Gama to Calecut.

Gonçalo Pirez, a mariner and retainer of Vasco da Gama. On May 31st 1497, he had been appointed master of a caravel recently built at Oporto.

Leonardo Ribeyro. According to Manuel Correa's commentary on the Obras do grande Camões, Lisbon, 1720, this, on the authority of the poet himself, is the full name of the "Leonardo" mentioned there.

João de Setubal, according to Correa, was one of the thirteen who accompanied Vasco da Gama to Calecut.

Alvaro Velho, a soldier. Perhaps this is the Alvaro Velho de Barreyro mentioned by Valentin Ferdinand (Valentino de Moravia or Alemão), in his Description of Africa in 1507, as having resided eight years at Sierra Leone.

Fernão Veloso, a soldier.

Convicts or banished men (Degradados) :

Pedro Dias, nicknamed "Northeasterling". Correa says that Vasco da Gama left him behind at Mozambique, and that subsequently he came to India.

Pero Esteves. Correa says that Vasco da Gama left him behind at Quiloa, and that when J. da Nova reached that port in 1501, he came out to meet him. Barros says that the convict who met J. da Nova had been landed by Pedro Cabral, and that his name was Antonio Fernandez.

João Machado, according to Correa was left behind by Vasco da Gama at Mozambique, but according to Barros it was Cabral who left him at Melinde, with instructions to make enquiries about Prester John. Cabral may have transferred him from Mozambique to the more northern port. He subsequently did good service, and Affonso de Albuquerque appointed him alcaide mór of Goa. He was slain in battle in 1515.

Damião Rodriguez was a friend of João Machado, and a seaman on board the São Gabriel, from which vessel he deserted at the shoals of São Rafael. When Cabral came to Mozambique, his grave was pointed out. All this is stated on the sole authority of Correa.

João Nunez, a "new" Christian (a converted Jew), who knew a little Arabic and Hebrew, and was landed at Calecut. In the Portuguese edition of Correa he is erroneously called João Martins.

Natives and others embarked in the east :

Gaspar da Gama. This is the "Moor", or renegade, who joined Vasco da Gama at Anjediva Island. Our anonymous author describes him as about forty years of age, and as being able to speak "Venetian" well. He claimed to have come to India in early youth, and was at the time in the service of the Governor of Goa. Vasco da Gama carried him to Portugal, where he was baptised and received the name of Gaspar da Gama. In the Commentaries of Afonso d'Alboquerque he is frequently referred to as Gaspar da India. Correa usually refers to him as Gaspar da Gama, but also calls him Gaspar de las Indias, or Gaspar d'Almeida. King Manuel, in his letter to the Cardinal Protector, calls him a "Jew, who turned Christian, a merchant and lapidary". Sernigi held a conversation with him at Lisbon. He speaks of him as a Sclavonian Jew, born at Alexandria. According to the information given by Barros and Goes, the parents of Gaspar fled from Posen, in Poland, at the time when King Casimir cruelly persecuted the Jews, in about 1456. After a short residence in Palestine they removed to Alexandria, where Gaspar was born. He accompanied Cabral as interpreter. Vespucci met him on his homeward voyage at Cape Verde, and in his letter of June 4th 1501, he speaks highly of Gaspar's linguistic attainments, and refers to his extensive travels in Asia. Gaspar repeatedly accompanied Portuguese expeditions to India, and was last heard of in 1510. Goes says that King Manuel liked him, and appointed him a cavalier of his household. Correa describes this Gaspar as a Jew, who, "at the taking of Granada was a very young man, and who, having been driven from his country, passed through many lands.....on to India". However, as Granada was only taken in 1492, this is absurd.

Lunardo da Chá Masser, who came to Lisbon in 1504 as ambassador of the Signoria, in a letter written in about 1506, says that Gaspar married a Portuguese lady, and was granted a pension of 170 ducats annually, in recognition of the valuable information that he furnished respecting the Oriental trade.

Monçaide, who came on board Vasco da Gama's vessel at Calecut, is stated by Barros and Goes to have been a native of Tunis, who, in the time of King João II had done business with the Portuguese at Oran, and spoke Castilian. He accompanied Vasco da Gama to Portugal and was baptised. In King Manuel's letter to the Cardinal Protector he is referred to as a "Moor of Tunis". The author of the Roteiro calls him a "Moor of Tunis" whom the Moors of Calecut suspected of being a Christian and emissary of the King of Portugal. Correa says that he was a native of Seville, who, having been captured when five years old, turned Moslem, although "in his soul he was still a Christian". He generally refers to this man as "the Castilian", and says that his true name was Alonso Perez. Castanheda tells us that the Portuguese corrupted Monçaide into Bontaibo, a combination of the Portuguese bom, "good", and the Arabic tayyib, having the same meaning. Monçaide is probably a corruption of El Masud, the "happy-one".

Malemo Canaqua, or Cana, the pilot who guided Vasco da Gama from Melinde to Calecut. He was a native of Gujarat. Malemo stands for "muallim" or "mallim", "master" or "teacher", the usual native designation of the skipper of a vessel, whilst "Kanaka" designates the pilot"s caste.

Davane, of Cambay, said to have been taken out of a dhau to the south of Mozambique, to have agreed to accompany Vasco da Gama to Calecut as broker, and to have been ultimately discharged with good testimonials in November 1498 at Cananor, is only mentioned by Correa. No other historian knows anything about this mythical personage.

Baltasar, and the four other Moors, forcibly carried away from Calecut were taken back by Cabral, as was also the Ambassador of the King of Melinde.

Vasco da Gama originally detained eighteen “Moors”. He is stated in the “Journal” to have subsequently liberated six, and to have sent one with a letter to the Zamorin. This would leave eleven, not five. The number of those liberated must, therefore, have been twelve, and not six.

CHAPTER 10

THE VOYAGE TO CAPE VERDE ISLANDS

The King was at Montemór o novo when he despatched Vasco da Gama and his fellow commanders upon the momentous expedition that was to place Portugal for a time in the forefront of maritime and commercial powers. With them went a caravel commanded by Bartolomeu Dias, whose destination was São Jorge da Mina. He had been made captain of the mine there as a reward for his many services.

It was summer, and His Majesty did not, therefore, desert the beautiful hills of Monfurado for the stifling heat of the capital, in order that he might witness the embarkation of his "loyal vassal" who, on account of his proved valour and past services he had deemed worthy of the honourable distinction of being entrusted with the conduct of so important an enterprise.

When all was ready, Vasco da Gama and his three captains went down to the chapel of Our Lady of Bethlehem (Belem), which Prince Henry the Navigator had built for his mariners, on the right bank of the Tagus. They all kept vigil there during the night of Friday 7th July 1497. On the following morning, which was Saturday 8th July 1497, they started in solemn procession for the place of embarcation, Vasco and his officers leading the way, with lighted candles in their hands, while a body of priests and friars followed, chanting a Litany. A vast concourse had assembled on the mud-flats that then lined the estuary. They stood bare headed in the blazing July sunshine, murmuring the responses to the Litany, and moving with the procession as it wound slowly across the foreshore, down to that landing place that Barros calls "a beach of tears for those who depart, a land of delight for those who arrive".

As the procession halted beside the margin of the river, the whole multitude fell on their knees in silence, while the vicar of the chapel received a general confession and granted absolution to all who might lose their lives on the voyage. Then Vasco da Gama and his comrades took leave of the weeping crowd and were rowed out to their ships. The royal standard was hoisted at the maintop of the São Gabriel. The

Captain's scarlet pennant fluttered above her crow's-nest. Eager and excited sailors ran to weigh anchor and unfurl the sails, on each of which was painted the great red cross of the Order of Christ. Friends and kinsmen said their last farewells. The attendant flotilla of small boats sheered off, and, with a stern wind filling their canvas, the four ships dropped down the Tagus, outward bound upon the longest and, with one exception, the most momentous voyage that had ever been undertaken.

Off the Canaries, the flotilla lay-to awhile and fished. Wind and weather had hitherto kept fair, but soon afterwards a fog descended, so dense that the São Rafael parted from her consorts. It had been arranged that the flotilla should in such a case meet at São Thiago in the Cape Verde archipelago. Off the neighbouring island of Sal, the São Rafael fell in with the Berrio, the storeship, and Bartolomeu Dias' caravel. Before long the flagship was also sighted some leagues ahead. The whole fleet arrived safely at São Thiago on the 27th of July. There they took in wood, provisions and water, and bade farewell to Bartolomeu Dias. The little armada left São Thiago on August 3rd 1497, with the intention of going further than Bartolomeu Dias' previous discovery, which was the Cape of Good Hope on the southern tip of South Africa, before going on to India.

Alvaro Velho, a Portuguese soldier who participated in the voyage, described in detail the voyage of Vasco da Gama in his "Roteiro da Primeira Viagem de Vasco da Gama" (The Diary of the First Voyage of Vasco Da Gama). It includes that part of voyage involving Bartolomeu Dias :

"In the name of God. Amen !

In the year 1497 King Dom Manuel, the first of that name in Portugal, despatched four vessels to make discoveries and go in search of spices. Vasco da Gama was the captain-major of these vessels; Paulo da Gama, his brother, commanded one of them, and Nicolau Coelho another.

We left Restello on Saturday July 8th 1497. May God our Lord permit us to accomplish this voyage in his service. Amen !

On the following Saturday [July 15] we sighted the Canaries, and in the night passed to the lee of Lançarote. During the following night, at break of day [July 16] we made the Terra Alta, where we fished for a couple of hours, and in the evening, at dusk, we were off the Rio do Ouro.

The fog during the night grew so dense that Paulo da Gama lost sight of the captain-major, and when day broke [July 17] we saw neither him nor the other vessels. We therefore made sail for the Cape Verde islands, as we had been instructed to do in case of becoming separated.

On the following Saturday, [July 22], at break of day, we sighted the Ilha do Sal, and an hour afterwards discovered three vessels, which turned out to be the storeship, and the vessels commanded by Nicolau Coelho and Bartholameu Diz [Bartolomeu Dias], the last of whom sailed in our company as far as the Mine. They, too, had lost sight of the captain-major. Having joined company we pursued our route, but the wind fell, and we were becalmed until Wednesday [July 26].

At ten o'clock on that day we sighted the captain-major, about five leagues ahead of us, and having got speech with him in the evening we gave expression to our joy by many times firing off our bombards and sounding the trumpets.

On Thursday August 3, we left in an easterly direction. On August 18, when about 200 leagues from Samtiaguo, going south, the captain-major's main yard broke, and we lay to under foresail and lower mainsail for two days and a night."

Having successfully prepared the voyage of Vasco da Gama, and having accompanied it to the Cape Verde Islands, Bartolomeu Dias, set off for São Jorge da Mina, in what is now Ghana, to take up his role as captain of the gold mine there.

Meanwhile, Vasco da Gama was to take the discovery of Bartolomeu Dias further, by rounding the Africa on his way to India.

CHAPTER 11

THE VOYAGE TO INDIA

The accounts of Vasco da Gama's remarkable voyage across the Southern Atlantic are of so scanty a nature that it is quite impossible to lay down his track with certainty. What we learn from the "Journal" may be condensed into a few words. The little armada left São Thiago on 3rd August 1497. Vasco da Gama stood south-east and parallel to the African coast until, in about 10° N., he reached the region of calms and encountered evil weather. When the wind blew at all, its direction was contrary. Sudden squalls arose from time to time and rushed down upon the fleet with tropical fury.

On the 18th August, when 200 leagues (680 miles) out at sea, the mainyard of the flagship sprung in a squall. This necessitated laying to for a couple of days and a night. To escape at once from the doldrums and from the baffling winds and currents of the Gulf of Guinea, Vasco da Gama conceived the bold and original idea of fetching a wide compass through the South Atlantic, so that, if possible, he might reach the Cape of Good Hope after circling round the tract in which the experience of Diogo Cão and Bartolomeu Dias had shown that unfavourable weather might be expected. He crossed the equator in about 19° W. and steered south-westward into an unknown Ocean.

In September, the fleet reached its westernmost limit, within 600 miles of South America, and then headed round for the Cape, beating slowly back against the south-east trades until, in higher latitudes, a west wind arose and carried the explorers on their way.

On November 4th, at 9am, the mainland of Africa was sighted, probably about 150 miles to the north of St. Helena Bay (30° S.), in what is now the Western Cape of South Africa.

On the 7th of November they dropped anchor in an inlet to which the commander-in-chief gave its present name of St Helena Bay. Since leaving the Cape Verdes his fleet had spent ninety-six days in the South Atlantic and had sailed fully 4,500 miles. No navigator of whom there is any authentic record had ever completed so long a voyage without sight of land.

At St Helena Bay, Vasco da Gama went ashore to take the altitude of the sun. On board it was impossible to obtain an accurate reading from his primitive astrolabes, owing to the motion of the waves. Here the ships were careened and fresh supplies of wood and water taken in. Here, too, the Portuguese made the acquaintance of some beach ranger Hottentots from a neighbouring kraal.

A soldier named Fernão Velloso received permission to accompany them to their kraal. On the way he was regaled with a banquet of roots and roasted seal, but some misunderstanding arose, and presently his comrades noticed him hurrying back with shouts and excited gestures. The Roteiro, however, states that Velloso's cries were heard on board the ships and a rescue-party set off at once. A scuffle ensued, in which the Hottentots bombarded the Portuguese with stones and arrows. A fish-spear struck Vasco da Gama in the leg.

No lives were lost in the skirmish and the voyage was resumed on Thursday, November the 16th. On Saturday the dreaded promontory came in sight. Legend had already surrounded it with fantastic perils similar to those that had made the whole Atlantic so formidable to medieval mariners. However, among Vasco da Gama's crews were seasoned men, who had served with Bartolomeu Dias, and it is unlikely that they believed the rumours current on shore. During four days, the wind was dead ahead, and it was not until noon on Wednesday, the 22nd of November 1497, that the Cape of Good Hope was finally doubled.

A few days later, the fleet cast anchor in the Angra de São Braz - the Bahia dos Vaqueiros of Bartolomeu Dias, the Mossel Bay of modern maps. Here, during a stay of thirteen days, the storeship was broken up and burned, and her contents transferred to the other vessels. Bartolomeu Dias had found the natives hostile at this point. They had stoned him when he came ashore to fetch water, and he had killed a native with a bolt from his crossbow. So Vasco da Gama was surprised on landing with an armed bodyguard, to find himself welcome.

On Friday the 8th of December, Vasco da Gama set sail once more, and by the 16th he had passed the Rio de Infante, the furthest landmark discovered by Bartolomeu Dias. Northerly winds, and the race of the Agulhas current, which here sets strongly inshore, flowing in a south-westerly direction, now carried the flotilla back, and at one point the pilots found themselves no less than sixty leagues abaft their dead reckoning. For a time it was feared that no further progress could be achieved, but a brisk wind sprang up astern, and presently the fleet came abreast of a land to which, as it was Christmas Day, Vasco da Gama gave the name of Natal.

Soon afterwards, they stood away from the land, either hoping to escape the force of the Agulhas current and to make northing in calmer seas, or fearing to be driven on a lee shore by the strong easterly wind. However, the mainmast of Paulo da Gama's ship was broken and an anchor lost through the parting of a cable. Finally the supply of drinking water ran short, so that it became necessary to cook with brine. The fleet was compelled to turn eastward again, and put in at the estuary of the Limpopo.

A crowd of Bantu, both men and women, had gathered on shore, and an interpreter named Martini Affonso, who had lived long in Congo and had learned some of the Bantu dialects of the west coast, was sent ashore with one of his comrades. They were hospitably received. All night long, crowds of men and women came to stare at the oddly coloured strangers.

On 16th January 1498, he continued his voyage and soon afterwards rounded the headland known, from the force with which the Agulhas current sweeps past it, as Cape Correntes. By so doing, he had unawares re-entered the civilised world.

Passing Sofala, as he had already passed Delagoa Bay, without sighting it, Vasco da Gama anchored in the estuary of the Kiliman or Quilimane River on the 24th of January. Thirty-two days (January 24th to February 24th inclusive) were spent in taking in water, cleaning the ships again, and repairing the broken mast. Just as all seemed to promise well, an epidemic of scurvy broke out among the sailors.

Men's hands and feet swelled, and their gums grew over their teeth so that they could not eat.

Mozambique, the next halting-place, was reached on the 2nd of March. It was a low-lying coral island, in the mouth of an inlet that afforded good anchorage.

The Shaikh, who Barros calls Zacoeja (possibly a corruption of Shah Khwaja), who governed Mozambique for his suzerain, the Sultan of Kilwa, exchanged courtesies with Nicolau Coelho and Vasco da Gama, and promised to furnish two pilots. On Saturday, March the l0th, the flotilla changed its quarters, taking up a berth off the adjacent island of São Jorge, where mass was said on Sunday. The Muhammadans now learned that their guests were Christians. So shocked were the pilots at this discovery, that one of them could only be retained on board by force. When the Portuguese attempted to seize the other, they were ordered off by a crowd of armed men in boats.

Leaving Mozambique on the 13th of March, the fleet lay becalmed for two days, and after drifting some leagues to the southward, was compelled to return to its moorings. The Shaikh sent a conciliatory message, but when the Portuguese attempted to disembark on the mainland for water, their landing was challenged, and during the night the Muslims erected a palisade for the defence of their springs. Vasco da Gama ordered the ships' boats to be launched, and artillery to be placed in the bows. His men kept up a cannonade for three hours and succeeded in killing two of the defenders, after which they rowed back in triumph to dinner.

Sufficient water was ultimately obtained, but as the wind remained light it was not until the 29th of March that the fleet was able to leave the neighbourhood of Mozambique, having secured two Arab pilots - skilled men, accustomed to the use of compass, quadrant, and navigating chart.

Mombasa was reached on the eve of Palm Sunday, April 7th. The Portuguese were eager to go ashore and join the supposed Christian community in the celebration of mass. After the usual exchange of presents between Vasco da Gama and the local ruler, two men were sent ashore and taken round the city. "They stopped on their way", says the Roteiro, "at the house of two Christian merchants, who showed them a paper, an object of their adoration, on which was a sketch of the Holy Ghost". Meanwhile, the orthodox folk of Mombasa were, in their turn, scandalised to learn from the Mozambique pilots and some half-castes captured by Paulo da Gama that the newcomers were dogs of Christians. Vasco da Gama's suspicions were soon aroused, especially when his pilots leapt overboard and escaped to a native dhow. By a judicious application of boiling oil, he induced two of the half-caste prisoners to divulge the details of a scheme by which his fleet was to be boarded and seized.

A daring attack was made at midnight by armed swimmers who strove to cut the anchor-cables of the Berrio and São Rafael. The watch aboard the Berrio at first mistook the splashing and movement for a shoal of tunny, but soon discovered their mistake and raised an alarm. Some of the Muhammadans had contrived to secure a footing in the mizzen-chains of the São Rafael, and were beginning to clamber up the shrouds; but on finding themselves detected they slid silently back into the water and vanished.

Despite these alarms, Vasco da Gama remained two more days off Mombasa, either, as Castanheda suggests, in the hope of securing a pilot, or because all the sick on board had benefited by the climate. He left on the 13th of April, still steering in a northerly direction, and at sunset on the following day cast anchor off Malindi.

An old "Moor" who with some companions had been captured in a dugout (almadia) during the brief run from Mombasa, was sent ashore to greet the Raja and assure him of Vasco da Gama's goodwill. The answer, accompanied by a gift of three sheep, was that the strangers might enter the port. They were free of all it contained, including the pilots. Not to be outdone in munificence, Vasco sent the Raja a return presents consisting of a cassock, two strings of coral, three wash-hand basins, a hat, some little bells and two lengths of striped cotton cloth. This did not end the competition : the Raja doubled his gift of sheep and added a really valuable consignment. of spices, while Vasco released all his Muhammadan captives. He had learned caution from experience. So on being invited to the royal palace, he answered with more discretion than accuracy that the King of Portugal had forbidden him to go ashore. He dared not disobey, so would the Raja honour him with a visit on board ?

To this the Raja replied, "What would my own subjects have to say if I ventured ?" But curiosity overcame fear, and he was rowed out to the ships, royally attired in a damask robe trimmed with green satin, and an embroidered turban. His dignity required the support of two cushioned chairs of bronze. A crimson satin umbrella protected him from the sun, and a band discoursed more or less sweet music on various kinds of trumpets, including two which were fashioned of ivory and were as large as the musicians who performed on them. In this state the Raja made a circuit of the Portuguese ships, while the artillery fired off salvos in his honour.

Nine days, from the 15th to the 23rd of April 1498, were spent in festivals, music and sham fights. On the 24th the Portuguese weighed anchor and set sail, under the guidance of a Gujarati pilot named Cana, steering east north-east across the Arabian Sea so as to fetch up at Calicut on the Malabar Coast of India.

For twenty-three days the ships held on a straight course, favoured by a steady breeze from the southwest, the herald of the winter rains. During three weeks no land was visible, but on Friday the 18th of May, after skirting the northern islets of the Laccadive group, the pilot gave orders to turn eastward, and presently the lookout signalled land ahead. It may be supposed - though the Roteiro and the chronicles are almost silent on the emotions aroused in this moment of supreme triumph - that all hands rushed on deck for a first glimpse of unknown Asia. They could discern, far away, the outline of mountain peaks rising above the horizon, dark under gathering clouds. On the morrow a thunderstorm broke, and a downpour of tropical rain blotted out the Malabar Coast before the pilot had time to take his bearings, but the heights that had first come into view may almost certainly be identified not as the main rampart of the Western Ghats, but as the outlying promontories of Mount Dely.

On the 21st May 1498, after a voyage lasting ten months and two weeks, the Captain-Major brought his ships to anchor off Calicut.

The ruler of the city was called by the Portuguese the Camorij, which is usually anglicised as Samuri or Zamorin. The origin of this term is doubtful. It may be a corruption of the Malayalam Tamuri (Sanskrit Samundri) Raja meaning "lord of the sea", and either a title or a family name. The Samuri was a Hindu who lived in a stone palace outside the city surrounded by his aristocracy - priestly Brahmans, and polyandrous Nairs or members of the fighting caste. Within the city Hindu mechanics and retail traders dwelt in wooden houses thatched with palm-leaves. A few stone buildings, including at least two mosques, had also been built by the rich Mopla merchants, descendants of Arab fathers and native women, who enjoyed a monopoly of maritime commerce.

Calicut was a free port, its ruler a Hindu bound by immemorial custom, which he could only disregard at his peril. He was probably not over eager for friendship with a crew of piratical adventurers whose exploits on the African coast may already have been denounced by the Moplas. His European guests told him of a mighty kingdom, so far away that its existence could not easily be verified. They sought an alliance and commercial privileges, but their only visible resources were three battered ships. Their gifts were unworthy of acceptance. Their very touch meant ceremonial defilement. Nevertheless, the Samuri was prepared to receive them with courtesy, at the risk of offending his best customers, the Moplas. As the monsoon was now at its height, and Vasco da Gama would not venture his ships inside the harbour for fear of treachery, the Samuri sent one of his own pilots to take them to a safe berth near Pandarani Kollam, some fifteen miles farther north. He also acceded to the request for an audience, and on the 28th of May Vasco da Gama landed with thirteen companions and started for Calicut in a palanquin carried by relays of bearers.

On the way he was taken by his native guides to a large stone pagoda, roofed with tiles, at the entrance of which rose a bronze pil1ar, tall as a mast and surmounted by the figure of a cock. Within was a sanctuary or chapel, containing a small image, which the guides were supposed to identify as a figure of the Virgin Mary. The Portuguese felt that their hopes were at last near fulfilment, and that the great discovery of a new Christendom in Asia had been achieved.

They all knelt in prayer, while the Hindus prostrated themselves and, if Damião de Goes can be believed, pointed to the image, crying "Maria, Maria !" The author of the Roteiro observes, without a trace of surprise, that "many other saints were painted on the walls, wearing crowns. They were painted variously with teeth protruding an inch from the mouth and four or five arms". The unorthodox aspect of these frescoes may have caused some misgiving, for Castanheda relates that one João de Sá, clerk aboard the São Rafael, exclaimed as he fell on his knees, "If these be devils, I worship the true God." The Roteiro adds that the qua fees who ministered in the church wore certain threads, "in the same manner as our deacons wear the stole" - an obvious allusion to the janeo or sacred cord worn by Brahman priests. "They asperged us with holy water," it continues, "and gave us some white earth that the Christians of this country are wont to sprinkle on the forehead and chest, round the neck and on the forearm." The Captain-Major, on receiving a supply of "white earth" for his private use, handed it to somebody else, giving the priests to understand that he would "put it on later". He may have discerned the fact that the principal ingredients of the sacred mixture were dust and cow-dung.

Arrived at Calicut, Vasco da Gama and his men were met by a native magnate, whose attendants escorted them through the town, marching to the lively strains of drums, trumpets and bagpipes, while every roof and window was thronged with spectators. After a scuffle at the palace gates, in which knives were unsheathed and several men injured - possibly owing to the pressure of the crowd - the Portuguese were ushered into the royal presence. They saw the Samuri reclining on a green velvet couch under a gilt canopy, and holding a massive golden spittoon in his left hand, while a cup bearer served him with betel from a golden bowl, so large that a man could hardly encircle it with both arms. After listening graciously to Vasco da Gama's recital of the virtues and resources of King Manoel, the Samuri replied that the ambassadors were welcome, and that he would regard their sovereign as a brother.

Custom required that all gifts should he forwarded to the Samuri through his factor and wali, who were summoned on the following day to inspect King Manoel's present. To the astonishment of these officials it comprised such articles as washing-basins, casks of oil and strings of coral - goods that might be acceptable to the headman of a savage African tribe, but seemed hardly an appropriate gift to the ruler of the greatest commercial port on the west coast of India. The factor and the wali were unable to conceal their amusement, and although Vasco da Gama sought an escape from his embarrassment by protesting that the gift came from himself, not from his sovereign, they advised him to send gold or nothing. As, however, gold was none too plentiful aboard his ships, the second audience proved final. The Samuri pertinently inquired what they had come to discover, stones or men. If men, why had they brought no gift ? But Vasco da Gama was allowed to present the letters he had brought from King Manoel, which were read aloud by Arab interpreters. He was granted liberty to land his goods and to sell them if he could find a purchaser.

On the 31st of May the Portuguese started back to Pandarani. The sun had already set, and as it was a windy night the native boatmen refused to undertake the long row out to the ships, which had been moored far from the shore. It was not until the 2nd of June that Vasco da Gama and his men were able to return on board. Meanwhile their suspicion had magnified the delay into an imprisonment. This was perhaps natural, as they were watched at night by armed guards - a precaution almost certainly intended to secure them from molestation by the Muhammadan traders, who spat ostentatiously whenever they met a Portuguese. Once he was safe on board, the Captain Major seems to have discarded his fears. He unloaded some of his merchandise and endeavoured to sell it, but the Muslim traders came only to scoff. Vasco da Gama then despatched a letter of protest to the Samuri, who replied courteously, sent an agent to assist in selling the goods, and finally had them conveyed at his own expense to Calicut.

From the last week of June until the middle of August the fleet remained off Pandarani. Meanwhile small parties of sailors went ashore and busied themselves in hawking shirts, bracelets and other articles. Their object was to raise enough money to buy samples of spices and precious stones. Their own goods, however, could not be sold except at a heavy loss, and at last Vasco da Gama sent Diogo Dias with a gift of "amber, corals and many other things" - to inform the Samuri that his ships were about to leave for Portugal, and to ask for a consignment of spices on behalf of King Manoel.

The Samuri's factor then explained that before the Portuguese departed they must pay the usual customs dues on the merchandise they had landed, amounting to 600 xerafins. The goods had been warehoused in Calicut, and left there in charge of a Portuguese factor, a clerk and a party of sailors.

A guard was set over these men, who were evidently to be held as hostages until the duty was paid. Vasco da Gama retaliated by seizing eighteen Hindus who had come to visit his ships. Among them were six Nairs whom it was necessary to exchange every day for other hostages. They would have starved to death rather than taste the "unclean" food provided by their captors.

On the 25th the ships stood off and anchored outside Calicut. They were presently joined by Diogo Dias, who brought a letter from the Samuri to King Manoel, written with an iron pen on a palm leaf. Its tenor, says the Roteiro, was as follows :- "Vasco da Gama, a gentleman of your household, came to my country, whereat I was much pleased. My country is rich in cinnamon, cloves, ginger, pepper, and precious stones. That which I ask of you in exchange is gold, silver, corals, and scarlet cloth."

Soon afterwards an exchange of hostages was effected : all the Portuguese and a portion of their merchandise were restored, the residue being probably withheld in lieu of duty. Vasco da Gama yielded up all his captives except five, whom he may have kept to compensate himself for the partial loss of his goods, though the Roteiro asserts that his object was to use these men "for the establishment of friendly relations" when he should return to India on a second voyage.

On Wednesday, the 29th of August, the Portuguese captains unanimously agreed that as they had discovered Christian India, with its spices and precious stones, it would be well to depart, especially as the Christians did not appear anxious for friendly interaction. That same day the ships set sail for Portugal.

The passage of the Arabian Sea was delayed by calms and contrary winds, while a terrible outbreak of scurvy caused the loss of thirty lives. So many of the crews fell ill that only six or seven men were left to work each vessel. The African coast was at last sighted, and on the 7th of January 1499 the fleet anchored once more in the friendly harbour of Malindi. But here also many of the crew died, and after five days the voyage was resumed. The São Rafael was abandoned and set on fire near Mozambique, because there were insufficient hands to work her, and the two surviving ships rounded the Cape of Good Hope on the 20th of March. About a month later they parted company, Nicolau Coelho taking the Berrio on to Lisbon, where he arrived on the l0th of July, while Vasco da Gama steered for the Azores in his flagship. There, in the island of Terceira, his brother Paulo died of consumption.

The date of Vasco da Gama's return to Lisbon is not certain, but it seems probable that he landed at Belem on the 8th or 9th of September 1499, and made his triumphal entry into the capital on the 18th September, the interval being spent in mourning and memorial masses for his brother.

The whole voyage had lasted some two years, and only fifty-five men returned out of the 170 who had sailed from Belem. However, the quest for Christians and spices had been accomplished. Because of the voyages of Bartolomeu Dias and Vasco da Gama, Portugal had become mistress of the sea-route to India.

CHAPTER 12

THE VOYAGE OF PEDRO CABRAL

When Vasco da Gama returned to Lisbon from India in 1499, he was received with great honour. He was made Count of Vidigueira and given an irrevocable commission to act as the chief of any future fleet to India, should he so desire.

Alvaro conceding to Dom Vasco da Gama the chief captaincy of all the ships departing for India during his lifetime, the king not being able to intervene in this matter, etc. :

"We, the King, make known to all to whom this our alvará may come, that in consideration of the very great and signal service that Dom Vasco da Gama of our council did to us and to our kingdoms in the discovery of India, for which reason we should give him all honour, increase and reward, and because of this, it pleases us that we grant him by this present alvará that of all the armadas that in his life we shall order made and shall make for the said parts of India, whether they be only for the trade in merchandise or whether it is necessary to make war with them, he may take and takes the chief captaincy of these, so that in the said armadas he has to go in person, and in them to serve us, and when he thus wishes to take the said captaincy, we may not place in them nor appoint another chief captain save him, because of his honour, and we confide in him that he will know very well our service; it pleases us that we grant and we in fact grant this reward and privilege as is said. Furthermore, we order to be given to him this our alvará by us signed, that we order shall be in every way kept and guarded, as in it is contained our reward, without impediment or any embargo that might be placed upon it. And it pleases us, and we wish that it be as valid as a letter by us signed and sealed with our seal, and passed by our Chancellery, in spite of our ordinance, even though it may not be passed by the officers of the Chancellery. Done."

To the Portuguese people Vasco da Gama's voyage was accomplished by the will of God, who had destined them for the control of the East, and, regardless of obstacles, they must continue. So the preparations were made for the dispatch of a second fleet to Calicut the following March.

The second expedition was of a less purely commercial character than the first, for the Rajah of Calicut, if he should prove contumacious, was to be punished and irreconcileable Arabs to be brought into subjection. The desire of the Portuguese to hasten these preparations was partly to prevent the Arabs from arming for defence and still further inciting the Hindus against them.

However, Vasco da Gama was tired. The voyage had been long and difficult, and he wished leisure to recuperate. It was Vasco da Gama's wish and that of the king that the leader of the next expedition to the East should be a man of a different type, who might be able to change to friendship the hostility that the native rulers had shown towards the Portuguese during the previous voyage. Perhaps Vasco da Gama had in mind also that after a more successful voyage by Pedro Cabral he himself might again return to show the Zamorin the true position of his country. At this time Vasco da Gama and Pedro Cabral were friends, and Vasco da Gama is said to have suggested his name for this office.

There were other reasons that induced Dom Manuel to select Pedro Cabral as chief captain of the Indian fleet. He had undoubtedly known him well at court. The standing of the Cabral family, their unquestioned loyalty to the Crown, the personal appearance of Cabral, and the ability that he had shown at court and in the council were important factors. Two of his brothers, João Fernandes Cabral and Luiz Álvares Cabral, were members of the council of Dom Manuel in 1499, and may have had some influence in this selection before the return of Vasco da Gama. The fact that Cabral was a collateral descendant of Gonçalo Velho, the honoured navigator to the Azores, may also have added a sentimental reason. The conditions that existed at court in those days are not recorded, but we know that there was much intrigue and jealousy. Cabral may have belonged to a faction that aided his choice. The selection of the chief captain for this fleet required great care. Cabral, therefore, must have been a man who was not only acceptable to Vasco da Gama and to the king but who also had the confidence of the people of Portugal and the respect of those who went with him.

Pedro Álvares Cabral was two years older than Dom Manuel, and was thirty-two years of age when he was selected in 1499 as chief captain of the fleet that was to go to India the following year. The name used in his appointment as chief commander of the fleet for India was Pedralvarez de Gouveia. Bartolomeu Dias was to accompany the fleet throughout its voyage.

The shipyards and arsenals were busy making ready one of the largest and certainly the most imposing fleets that had hitherto sailed the high seas of the Atlantic. The purpose of the voyage meant that the ships constituted an armada as well as a flotilla. So it is probable that some of the vessels were equipped with heavy armament such as bombardas and culverins. Provisions and supplies for twelve hundred men for a year and a half had to be provided, and a cargo with which to trade.

The selection of the officers and crew was made with great care. There was no difficulty in securing them. They were to go with pay and not subject to the reward of the king as were those of Vasco da Gama. The captains were chosen to impress the rulers of India with the greatness of Portugal, and for this reason members of noble families commanded many of the ships. There went also Franciscan friars and clergy, some of whom were to remain in India. The cargo was in the charge of a factor with assistants, for a factory was to be permanently established at Calicut.

CHAPTER 13

THE FLEET OF PEDRO CABRAL

An inscription on the South America of the Cantino Chart alleges that Pedro Cabral had fourteen vessels in his fleet. It runs thus : "A vera cruz + chamada p. nome aquall achou pedraluares cabrall fidalgo da cassa del Rey de portugall & elle a descobrio indo por capita moor de quatorze naos que o dito Rey mandaua a caliqut en el caminho indo topou com esta terra aqual terra se cree ser terra firma em aqual a muyta gente de discricam & andam nu os omes & molheres como suas mais os paria[m] sam mais brancos que bacos tem os cabellos muyto corredios foy descoberta esta dita terra em aera de quinhentos."

Cabral's fleet consisted of both ships and caravels. There is no official document that tells how many belonged to each class, and the only authors who give us exact statements are Castanheda, who says that there were three round ships, and the rest were ships (probably meaning caravels), and Gaspar Corrêa, who states there were ten large ships of 200 to 300 tons and three small ones. In the account of the voyage written by a Portuguese pilot and published by Ramusio, Pedro Cabral is said to have commanded thirteen vessels. Barros also avers that the fleet comprised thirteen sail. Castanheda, who gives the best narrative of the voyage and usually refers to the vessels as caravels, was probably right. Corrêa had the classes reversed. This uncertainty in the description of the vessels is due to the inexact way in which ships were then designated. The Capitania, or flagship, and the ships of Simão de Miranda and Sancho de Tovar were undoubtedly naos redondos, or round ships. The remaining vessels were probably small ships and caravels, with possibly a few caravelas redondas that combined the two types. There is no description of any of the ships of Cabral's fleet. The illustration of the fleet shown in the Livro das Armadas da India was made long after the voyage, and was derived from references to the ships as given by the historians. It is therefore of little real value. The ships must have been similar to those in Vasco da Gama's fleet and to other vessels of the early sixteenth century.

The "round ships" were so called because when viewed from the front or rear they appeared round on account of their wide beam and bulging sails and to distinguish them from the "long ships" or galleys of the Venetians. These round ships were provided with castles fore and aft, which were used as living-quarters, and which were also of advantage for boarding in case of a naval engagement. They had three masts. The foremasts and mainmasts were square-rigged, and the mizzen-mast had a lateen sail. There was also a square sprit-sail at the bow. No sails were employed above the top sails, but in fair weather bonnets were used. The caravels had three or four lateen-rigged masts and were often provided only with an aft castle. With the caravela redonda the foremast was square-rigged and the others lateen-rigged. This type had the advantage of being steadier than the caravels and of permitting the use of two castles. It is doubtful if any of the ships in the fleet exceeded 300 tons, and the smallest was not over 100 tons. The ships of Pedro Cabral and Simão de Miranda were the largest and may have had a capacity of 250 or 300 tons. Because Sancho de Tovar went as second in command, his ship of 200 tons would probably only be exceeded in size by these two. A comparison of the respective tonnages of the fleets of Vasco da Gama and Pedro Cabral shows that Cabral's fleet was about five times the size of Vasco da Gama's. Six of Cabral's vessels were lost at sea. From a financial standpoint, that of Sancho de Tovar was the most serious.

Some of the vessels of the second Indian fleet may have been merchantmen freighted by merchants or mercantile firms and permitted by the king to accompany the fleet under conditions such as are preserved for us by João de Barros who writes in his relation of the voyage of João da Nova to India in 1501 with four ships after that of Cabral's voyage in 1500 : "The captains of the other ships" (than that of Da Nova himself) "were Dioga Barbosa, servant of Dom Alvaro, brother of the Duke of Brangança for the ship was his and Francisco de Novaes, servant of the king and the other captain was Ferñao Vinet, a Florentine by nation, because the ship in which he sailed belonged to Bartolomeo Marchioni, who was also a Florentine, a resident in Lisbon and the richest the city had produced at that period. In order that those of this kingdom who were engaged in commerce might have a trade open to them, the king ordained that they should be permitted to freight vessels for those parts, some of which have sailed and others are laded and this method of bringing spices is still employed. And, seeing that upon those persons to whom the king granted this concession it was imposed as a condition in their contracts that they must present for the approval of the king the captains of the ships or barques that they freighted and whose appointments were to be confirmed by the king, they often proposed posed persons selected rather because they were well fitted for the business of the voyage and the charge laid upon them than because they were of noble blood." Evidence that one of the vessels that accompanied the fleet commanded by Pedro Cabral was freighted by a Lisbon merchant is contained in a letter from D. Cretico, Envoy from Venice to the King of Portugal, dated 27th June 1501 and written to an unnamed correspondent who is addressed as "Serenissime princeps." "This ship that has arrived," writes the Envoy, "is that of Bartholomio, the Florentine, with the cargo, that consists of about 350 cantaros of pepper, 120 cantaros of cinnamon, 50 to 60 cantaros of lac and 15 cantaros of benzoin".

CHAPTER 14

THE CREW OF PEDRO CABRAL

There was no difficulty in securing men for Pedro Cabral's voyage. For this reason it was felt wise to determine their pay in advance. The king in consultation with Vasco da Gama, Pedraluarez Cabral and Jorge de Vasconcellos, superintendent of the royal store-houses, (Provedor dos almazens do Reyno) fixed the salaries and wages to be paid to the officers and men.

The Chief Captain was to receive ten thousand cruzados, five thousand of which were paid in advance. The captains, including Bartolomeu Dias, were to receive one thousand cruzados for every hundred "toneis" of their ships and one thousand cruzados of the total amount was to be paid in advance. One year's wages, one hundred and thirty cruzados, were to be paid in advance to the married able-bodied seamen, and sixty-five cruzados to the unmarried. Six months' wages, sixty-five cruzados, were to be paid in advance to the ordinary seamen, if married, and thirty-two cruzados and a half to those who were not married.

Gaspar Corrêa tells us that "What was decided was that the chief captain of the armada should have for the voyage 10,000 cruzados and 500 quintals of pepper, paid for from his salary of 10,000 cruzados at the price at which the king might purchase it, on which he should not pay taxes, except the tenth to God for the monastery of Nossa Senhora de Belem; and to the masters and pilots 500 cruzados for the voyage and thirty quintals of pepper and four chests free; and to the captains of the ships 1,000 cruzados for each 100 tons, and six chests free, and 50 quintals of pepper for the voyage; and to the mariners 10 cruzados per month and ten quintals of pepper for the voyage and a chest free; and to every two ordinary seamen, the same as one mariner; and to every three pages, the same as to one ordinary seaman; and to the mates and boatswains, as to a mariner and a half; and to the official men, that is, in each ship two caulkers, two carpenters and two rope makers, a steward, a bleeding barber, and two priests, the third of that of two mariners; and to the men at arms, five cruzados per month and three quintals of pepper for the voyage; and in each ship went a chief gunner and two bombardiers; to the chief gunner 200 cruzados and 10 quintals of pepper for the voyage and two chests free; and to the bombardiers the same as to mariners; and to each man at arms his free chest. And all the quintals of pepper loaded with their money with only the tenth to God; and the payment of this pepper to be made to them by the king in money, according to the price he might sell it for with a deduction, if any, because the pepper dried on the voyage, a soldo to a liura; and payment in advance to the men at arms, and one year in advance to those married, and to bachelors half, and the same to all officials of the ships, and to the chief captain, 5,000 cruzados, and to each captain 1,000 cruzados, and to the men at arms six months each, and in their chests white clothing."

The chief officers and pilots occupied the aft castle. On Cabral's ship provision was also made for the meetings of the council and for entertainment. The chief factor with his principal aids evidently went with Cabral and they too would be quartered there. The crew, each provided with a bed roll and a chest, slept below decks. In the waist of the ship, cannon could be placed on either side, and in the centre was a large hatchway into which the ship's boats were lowered. The caravels probably had a bombard at the bow. Sails were manipulated with winches or capstans, that were also used for handling the cargo. At the stern was hung the farol, an iron cage in which firewood was burned at night. The sombre pitch-covered hulls were relieved by the bright colours with which the superstructures were painted. The flagship determined the speed and the changes in the course of the fleet; the others followed.

The food for the crew consisted of biscuit, dried or salted meat and fish, rice, sardines, dried vegetables and fruits, particularly figs. Oil, honey, sugar, salt, and mustard were provided. Wine was evidently furnished to the crew, because large quantities of it were carried on other fleets. The officers naturally fared better. Caminha states in his letter that chickens and sheep were carried on the ships. The crew suffered greatly from scurvy, and oranges were obtained as a remedy whenever possible. In addition to large quantities of provisions and supplies the fleet had cargo for trade. Two of the caravels, those of

Bartolomeu Dias and Diogo Dias, were destined for the coast of Sofala. These carried copper and small wares, such as looking-glasses, bells, and coloured beads, which the Portuguese, in their trade on the Guinea coast, had learned were desired by the natives. More valuable cargo may also have been taken for trade with the Arabs. The main fleet took copper, in bars or worked, vermilion, cinnabar, mercury, amber, coral of various grades, and cloths, particularly fine woollens, satins, and velvets in bright colours. The latter were chiefly used by the rich for decorations, since the people of the East were satisfied with their scanty cotton garments. Silks could be obtained to better advantage from China and embroideries from Cambay. For their purchases in the East the Portuguese also carried gold. This was in currency. As this was desired because of its intrinsic value, coins of nations other than Portugal were also taken. The coins of Venice were particularly esteemed, because they were better known. These coins were called "trade money" and were those that had not been greatly debased. Most of the money was carried in the flagship, since the factors made purchases for the king, and every care must be taken that his interests be protected. The representatives of the Italian merchants and Ayres Correia with his staff were probably also provided independently, as they evidently traded on their own account, with a percentage deducted for the Crown.

The Captains

On comparing the names of the captains of the ships as we find them in the "Asia" of Barros with the names mentioned by Correa we find that Correa has ten of the thirteen names mentioned by Barros and four that are not amongst the thirteen. The number of captains is thus raised from thirteen to seventeen. According to Barros, Simão de Miranda sailed in a different vessel from the Chief Captain. The captains whom Barros mentions by name are the following : Pedraluarez Cabral; Sancho de Tovar, son of Martin Fernandez de Tovar; Simão de Miranda, son of Diogo de Azevedo; Aires Gomez da Silva, son of Pero da Silva; Vasco de Taide; Pedro de Ataíde, whose sobriquet was "Inferno"; Nicolao Coelho; Bartolomeu Dias; Pero Dias; Nuno Leitão; Gaspar de Lemos; Luis Pirez and Simão de Pina. The additional names to be found in the "Lendas da India" are those of Bras Matoso, Pedro de Figueiró, Diogo Dias and André Gonçalves. Three small vessels (navios pequenos) were commanded by Luis Pirez, Gaspar de Lemos and André Gonçalves.

The story, of the voyage as it is told by Barros and Correa will not permit us to omit any of the Captains whose names they record and the only possible reconciliation of the two narratives lies in the adoption of all the names as genuine. We must believe therefore that seventeen, or, if we include the ship belonging to Bartolomeu Florentym, eighteen vessels constituted the fleet that was led by Pedro Cabral.

The following is known of those captains :

Bartolomeu Dias the discoverer. He had three brothers, Pero, Alvaro, and Diogo. Pero accompanied him on his first expedition, and Diogo was a captain of Cabral's fleet.

Diogo Dias, a brother of Bartolomeu, who had gone with Vasco da Gama as a writer on the São Gabriel. On Cabral's voyage he was the captain of the caravel that, having become separated from the fleet in the South Atlantic, sailed too far east and discovered Madagascar. Caminha speaks of him in his letter as a jovial man who was well liked by his companions.

Sancho de Tovar, or Toar, sent with the fleet as second in command with powers to succeed Pedro Cabral in case of the latter's death. He was a Castilian fidalgo, who, after killing the judge who had condemned his father to death for following the side of Afonso V against Ferdinand and Isabella, fled to Portugal. His appointment as a member of the fleet was evidently due to his loyalty to the Portuguese Crown. The choice, however, does not seem to have been a happy one. His ship, probably the El Rei, of 200 tons, ran ashore near Malindi and was lost with its cargo of spices. Tovar later took command of the caravel of Nicoláu Coelho and visited Sofala. He did not sail again to India.

Simão de Miranda, a nobleman and son-in-law of Ayres Correia. His name is placed third in lists of captains by all authorities except

Castanheda. His ship was probably about the size of the flagship. Because it accompanied the flagship and was sent on no special missions, this ship and its commander are mentioned only incidentally by the historians. It may have contained merchandise belonging to Ayres Correia and other Portuguese officials that did not belong to the Crown. It evidently took on cargo at Calicut. Miranda died in 1512, when captain of Sofala.

Aires Gomes da Silva, a nobleman of highest rank. His caravel was lost during the storm in the South Atlantic.

Vasco de Ataíde, a nobleman. According to Caminha and in the first edition of Castanheda he commanded the ship that lost company near the Cape Verde Islands. Other authors state that this was commanded by Luis Pires. Neither reached India, so the question is not of importance. Since Caminha saw the captains often while in Brazil, his statement cannot be questioned. The author of the Anonymous Narrative, a contemporary document, states that Vasco de Ataíde did not return to Lisbon. Castanheda, and following him de Barros and de Goes, claim that he did, although later Castanheda says that six ships were lost, which evidently included this one. According to Corrêa his vessel was a poor sailor and had difficulty in keeping up with the rest of the fleet. There is also a divergence of opinion as to whether the ship was lost during a storm. Caminha, our best authority, says that it was during clear weather.

Pedro de Ataíde, a nobleman and probably related to Vasco de Ataíde. Vasco da Gama married Catarina de Ataíde after his return from India, and this may have influenced the selection of the two Ataídes as captains. It was the caravel of Pedro, the São Pedro, which was sent to secure the elephant for the Zamorin. His ship was loaded at Cranganore. On the return voyage it became separated, but rejoined the flagship at Beseguiche. Pedro de Ataíde went again to India with Vasco da Gama in the São Pedro. He accompanied the fleet of Sodré to the Straits and returning with Francisco de Almeida was shipwrecked and died at Mozambique. De Barros and de Goes give him the nickname of “Inferno”.

Nicoláu Coelho, an experienced captain who had gone with da Gama as commander of the Berrio and took an active part in that voyage. He returned to Lisbon before da Gama. Nicoláu Coelho was a fidalgo of great valour to whom El-Rey, Dom Manuel, gave the captaincy of a ship to go in company with the great Vasco da Gama to discover India, in which he acted with great distinction and prudence; and when he returned, he arrived first at Cascaes before Vasco da Gama. Through him the king learned of all that happened in that discovery. He again sailed to India with Cabral, possibly in the same caravel. On the return voyage he replaced Nuno Leitão da Cunha as the commander of the Anunciada and reached home nearly a month before the rest of the fleet. He went to India a third time in 1503 under Francisco de Albuquerque and on the return voyage was shipwrecked and died with that commander in January 1504.

Nuno Leitão da Cunha, whom de Barros calls a cavaleiro. He commanded the Anunciada, which was financed by Marchioni and other Italians. This was one of the smallest though the fastest of the caravels. It was this captain who saved the life of Antonio Correia, the son of Ayres Correia, at Calicut. Da Cunha filled an important position at Lisbon after his return.

Gaspar de Lemos, a fidalgo about whom little is known. He commanded the supply-ship that returned from Brazil carrying letters to the king. This was the vessel that could best be spared from the fleet. Nothing is known of its return voyage. No place names are recorded on subsequent maps to indicate that it skirted the coast to the north for the purpose of further discovery. The credit for the discovery of this coast probably belongs to Vespucci, who after landing at Cape Saint Roque followed it to the south. De Lemos probably proceeded direct to Lisbon, in accordance with Cabral's instructions.

Luis Pires, who may have been the captain of the caravel financed by the Count of Porta Alegra. Nothing is known of his life, and his ship capsized during the storm.

Simão de Pina, a nobleman who was related to the chronicler Ruy de Pina. He commanded a caravel that was lost during the storm.

The Factors

Three factors were identified with Cabral's fleet, Ayres Correia, the chief factor, Afonso Furtado, who was factor of the two caravels destined for Sofala, and Gonçalo Gil Barbosa, who went out as a writer but was left in charge of the factory at Cochin. There evidently were other assistant factors, some of whom were lost on the voyage, because according to the Instructions one factor was to go with each ship. It is probable that the Italian merchants had one of their own on the Anunciada. Corrêa names Gonçalo Gomes Ferreira as a factor who was left at Cananore, but he is not mentioned by other historians.

The duties of the factors were to make commercial treaties, to conduct trade with the natives, and to take charge of the cargo. The writers were under their supervision and to them their duties were sometimes delegated. While under the authority of the chief captain, the factors were largely independent and were governed by a special section of the Instructions. Their salary is not given by Corrêa and it seems probable that other arrangements were made for them, either in the form of commissions or permission to trade on their own account.

Ayres Correia was evidently an experienced merchant in Lisbon with a knowledge of Eastern commodities. According to Castanheda it was from him that the store-ship of two hundred tons in the fleet of da Gama was purchased. As chief factor he was an important member of the council but looked to Cabral as his superior officer. In addition to all matters connected with trade, he had in his charge the making of commercial treaties. On this account he may be considered almost on an equal footing with the chief captain. He spoke Arabic fluently and probably had previously traded in Morocco. Correia has been blamed for the massacre in Calicut. This was due largely to lack of knowledge of Malayalam and to over-reliance on the word of the Arab traders. He died fighting on the shore. His son, Antonio, a boy of twelve, who was saved, later became one of the most famous captains in the East.

Afonso Furtado is given by Castanheda as the factor who was to be left at Sofala. De Barros and de Goes state that he was to be left there as a writer. He probably filled both offices. Bartolomeu Dias and Diogo

Dias, with whom Furtado was to remain in Sofala, were to stay on the East African coast and carried cargo for that purpose. On this account Furtado was sent ashore at Kilwa, the capital of the coast of Sofala. He may have succeeded Ayres Correia after his death.

Gonçalo Gil Barbosa was a brother of Diogo Barbosa, who was in the service of the Duke of Bragança and in that of Dom Alvaro, who sent a ship with Cabral's fleet, but which was lost in the South Atlantic. It was probably through Dom Alvaro that Gonçalo Gil Barbosa was able to secure the position of writer under Ayres Correia. Diogo Barbosa had a son, Duarte Barbosa, the author of the Book of Duarte Barbosa, who it has been claimed accompanied his uncle and remained with him at Cochin. Gonçalo Gil Barbosa was acting as factor at Cochin and was left there when Cabral's fleet hurriedly departed for its return voyage. When da Gama reached India on his second voyage, Barbosa was transferred to Cananore to take charge of a permanent factory that was established there. Corrêa gives him the name of Gil Fernandez. He seems to have learned Malayalam while at Cochin and was thus of great value in the development of commercial relations both there and at Cananore.

The Writers

The writer (escrivão) or clerk kept the records and accounts and made the reports for the factors. We do not know how many accompanied the fleet, but probably at least one for each ship. From the fact that two were left at Cochin there seem to have been more than those whose names have been recorded. Pero Vaz de Caminha, Gonçalo Gil Barbosa, and João de Sá may have ranked above the others, and may because of this have had the duty of writing the account of the voyage. We have previously mentioned the two former.

João de Sá had gone with da Gama on his first voyage as a clerk on the São Rafael. He was held in high esteem by Vasco da Gama and when da Gama left for Terceira with his dying brother, Paulo, he was given the command of the São Gabriel. His name is among those who went with Pedro de Ataíde to capture the ship from Cochin. De Sá was later treasurer of the India House.

Other writers mentioned are Lourenço Moreno and Sebastião Alvares, who were left at Cochin, and Diego de Azevedo and Francisco Anriquez, who, Corrêa says, were selected for Calicut and Cananore. Corrêa gives Fernão Dinis in place of Sebastião Alvares.

The Pilots

Each of the ships seems to have had a pilot, though the office of pilot and master may have been combined in some of the smaller vessels. Six ships sailed independent courses during parts of the voyage. Only the flagship and that of Simão de Miranda remained continuously together.

The pilots were evidently under one or more chief pilots who remained on Cabral's ship. With the chief pilots were associated the native pilots for the East African coast and the Indian Ocean.

Caminha gives the names of two pilots, Afonso Lopez and Pero Escolar, but none is mentioned by other writers. Afonso Lopez is referred to by Caminba as 'our pilot', which may indicate that he was one of those with Cabral on the flagship.

Pero Escolar did good service with da Gama and was rewarded by the king on his return. He, too, may have been on Cabral's ship. Pero Escolar continued to act as pilot in succeeding fleets. In November 1509 he was at Cochin in this capacity, and in 1515 he was the pilot of the ship Conceyçam.

Brito Rebello believes that João de Lisboa, one of the most notable pilots at this time, also went with Cabral's fleet. It is not possible to state definitely that João de Lisboa went with Cabral, but it may be assumed that with Nicoláu Coelho would go one who was with him on the first voyage as Pero d'Alemquer, who went with Dias, followed in that of Gama.

Besides the pilots there also went in the fleet an astronomer, After the fleet left Brazil, Master John is not heard of again and so he may have continued on one of the smaller vessels that was lost.

The Interpreters

Only two official interpreters are mentioned as being with the fleet, Gaspar da Gama and Gonçalo Madeira of Tangiers, who Castanheda says was left at Cochin. There were others, however, who spoke Arabic. Ayres Correia seems to have had the best knowledge of that language. He probably knew the dialect spoken in Morocco and may have had some difficulty in speaking correctly the language used by the Arabs in India. In the relations of the Portuguese with the Indians, some Arab interpreters were necessary. Because of the lack of Portuguese interpreters, many misunderstandings arose at Calicut. The native fishermen whom da Gama had brought to Portugal had been taught Portuguese, but these, because of their low caste, were nearly useless. While Gaspar da Gama is called an interpreter, he does not seem to have been of much assistance in that capacity during Cabral's voyage, and does not appear to have been ashore at Calicut at the time of the massacre. He was apparently treated with some distrust in spite of his conversion and marriage in Portugal as told by Ca' Masser.

Other Members of the Fleet

Sancho de Tovar, as has been said, was the captain who went to succeed Cabral in case of his being incapacitated. Apparently there were other noblemen who went with similar instructions to replace other captains, or who filled subordinate positions. Several of these are mentioned. Among them were Dom João Tello, who is referred to by Caminha, and the Spaniard, Pedro Lopez de Padilla, who is named in the letter of Dom Manuel written in 1501.

Men of minor importance who have not been mentioned elsewhere are Vasco da Silveira (de Barros), Fernão Peixato and João Rois, who were saved by Coje Benquim at Calicut (Castanheda), Fernão Perez Pantaja (Corrêa), who accompanied Duarte Pacheco, and Vasco da Silveira on the caravel that attacked the ship from Cochin, and Francisco Correa (Osorio) and Diogo de Azevedo (Corrêa), who were sent by Cabral to the Zamorin. Gonçalo Peixato is also named as one who escaped after the attack at Calicut (Osorio).

The natives who went with the fleet were Baltasar and four Indian fishermen, whom Vasco da Gama had taken from Calicut by force, Moorish pilots who Corrêa says had returned with da Gama, and an ambassador from the King of Malindi.

On the homeward voyage there came the two Christians, Priest Joseph and Priest Mathias, two natives from Cochin, an ambassador from the King of Cananore, possibly the converted Indian yogi, Miguel, though this is not sure, and a hostage from Sofala.

Friars and Priests

Vasco da Gama had reported that the people of India were Christians though not using the rites of Rome. For their instruction in the Catholic Faith, Franciscan fathers, well educated in the doctrines of the Church and strict observers of its rites, were sent in Cabral's fleet. These have been identified as follows : Frei Henrique, of Coimbra, guardian, Frei Gaspar, Frei Francisco da Cruz, Frei Simão de Guimarães, Frei Luis do Salvador. All of these were preachers and theologians. There went in addition Frei Maffeu as organist, Frei Pedro Neto, a chorister, and Frei João da Vitoria, a lay brother. Frei Henrique had formerly been a judge of the Casa da Supplicação. He took the Franciscan habit in the convent of Alemquer and became celebrated for his learning and eloquence. Frei Henrique officiated at the first religious services in Brazil. The Franciscan brothers were on shore at Calicut when the factory was attacked, endeavouring to attend to their religious duties although they did not understand the native language. Three were lost during the massacre, and Frei Henrique, who was wounded, narrowly escaped. He returned to Portugal. He was Bishop of Ceuta in 1505, confessor to Dom Manuel, and Inquisitor. In the last position he presided at the first burning of a Jew in Portugal, at Lisbon. Frei Henrique died at Olivença in 1532. No record remains of the religious efforts of these fathers during the voyage, except the mention of the conversion of the yogi christened Miguel. There were also in the fleet eight priests, in the charge of a vicar, whose names are not known. The vicar as representing a bishop had jurisdiction in his behalf.

CHAPTER 15

VASCO DA GAMA'S MEMORANDUM

There is a memorandum supposed to have been furnished by Vasco da Gama, regarding the conduct of the fleet at sea before reaching the Cape of Good Hope.

Since Vasco da Gama had sailed to the Cape of Good Hope by the direct route, his advice to Cabral, who was to follow a similar course, would have been of assistance. Such advice was evidently secured and incorporated in Cabral's instructions.

Varnhagen, in his search for early documents in the Archives of Portugal, discovered the most important portion of Cabral's instructions. A short time later he found one leaf of a memorandum, apparently by Vasco da Gama, at a sale of old papers. This, he inserted in facsimile in the first edition of his Historia geral do Brasil. He there claimed that the document had been sent to the Torre do Tombo for preservation. However, there is no record that this was ever received, nor can it be found there. In spite of this cloud on its authenticity, the memorandum may still be accepted with some degree of confidence, for the instructions for later voyages resemble portions of it almost exactly.

It does not seem to have been written by Vasco da Gama but more probably, as Dr. Antonio Baião suggests, by the Secretary of State, Alcaçova Carneiro, during an interview with Vasco da Gama.

These notes were evidently those incorporated in the official instructions that Pedro Cabral probably issued to the captains of the various ships.

Vasco da Gama was impressed with the necessity for preventing the loss of convoy by the ships, since he probably had some difficulty in this respect during his voyage.

The methods he suggested were not new. Whenever ships went

together out of sight of land, similar methods must have been used, but these evidently varied, and the recommendations of Vasco da Gama were those to be adopted for this voyage.

This memorandum is important, for it shows that Vasco da Gama not only suggested that the fleet should proceed in a southerly direction from the Cape Verde Islands and then east to the Cape, but he also advised, if the winds were favourable, that it should continue to the south-west from those islands, a course that Pedro Cabral followed. Inasmuch as these directions were entirely for navigation, there are no indications that a divergence westward was intended for any other purpose.

The memorandum consisted of more than one page, as is indicated by the fact that the sentence at the end is incomplete. The text is written in the centre of the page with notes on either side. It is crossed with lines showing that the information had been used and embodied in a more carefully worded document.

The translation given is from the text in the História da Colonização Portuguesa do Brasil, vol. i, pp. Xvi-xix :

“This is the way that it appears to Vasco da Gama that Pedroalvarez should follow on his voyage, if it please Our Lord.

In the first place, before he departs from here, to make very good ordinance so that the ships will not be lost some from the others, in this manner : namely, whenever they are obliged to change their course, the chief captain shall make two fires, and all shall respond to him, each with two similar fires. And after they thus respond to him they shall all turn. And he shall thus have given them the signals : that one fire will be to proceed, and three to draw the bonnet and four to lower sails. And none shall turn or lower sails or draw bonnet unless the chief captain shall first make the aforesaid fires, and all have replied. And after sails shall thus have been lowered, none will be hoisted until after the chief captain makes three fires, and all have replied [A]. And if any is missing they shall not hoist sail but only go with lowered sails until the coming of day, so that the ships cannot be carried so far that they are unable to see one another by day. And any ship that has its rigging down will make many fires to summon other ships so that it may be put in order.

After in good time they depart from here, they will make their course straight to the island of Samtiago, and if at the time that they arrive there, they have sufficient water for four months, they need not stop at the said island, nor make any delay, but when they have the wind behind them make their way towards the south [B, C]. And if they must vary their course let it be in the south-west direction. And as soon as they meet with a light wind they should take a circular course until they put the Cape of Good Hope directly east. And from then on they are to navigate as the weather serves them, and they gain more, because when they are in the said parallel, with the aid of Our Lord, they will not lack weather with which they may round the aforesaid cape. And in this manner it appears to him that the navigation will be shortest and the ships more secure from worms, and in this way even the food will be kept better and the people will be healthier.

And if it happens, and may it please God that it will not, that any of these ships become separated from the captain, then it must sail as well as it can to make the Cape and go to the watering place of Sam Bras. And if it gets here before the captain it should anchor in a good position and wait for him, because it is necessary for the chief captain to go there to take on water so that henceforth he may have nothing to do with the land, but keep away from it until Mozambique for the health of his men, and because he has nothing to do on it. [D, E, F.]

And if it be the case that the chief captain comes first to this watering place, before the ship or ships that are lost from him...

[The following notes, that appear on either side of the text, are indicated where they seem most appropriate.]

A (left). Save that if one of the ships cannot stand the sail as well as the captain's ship, and the strength of the wind requires him to draw it.

B (right). If the ships on leaving this city, before they pass the

Canaries, should encounter a storm so that they have to return, they shall do everything possible so that all may return to this city. And if any one of them cannot do so, every effort should be made to reach Setuuel [Setubal]. And wherever it may be, it will at once make known here where it is, so as to receive orders as to what it should do.

C (left). They will return before the island of Sam Nicalao in case this is necessary [; or] because of sickness at the island of Samtiago.

D (left). If these ships departing from this coast should become separated from each other in a storm, so that some make for one port and others for another, the manner in which they are to join each other : and if the signs of guidance are not made by some one of the ships and it cannot be seen you will, with all the rest, make your way straight to the watering place of Sam Bras.

E (right). And there, while you take on water, the aforesaid ship will be able to overtake you. And if it does not overtake you, you will depart when you are ready, and will leave there for it such signs that it may know upon arriving there that you have gone on, and will follow you.

F (left). And signs should be set up, where routes are to be taken for the ships that lose each other, and this will be done with the very good experience of all the pilots."

CHAPTER 16

THE DISCOVERY OF BRAZIL

On the 8th of March 1500 the fleet of thirteen vessels, including Bartolomeu Dias as captain of one of the ships, was assembled in the Tagus, some three miles below Lisbon, near the small hermitage of Restello, where the monastery of the Jeronimos now stands.

Before their departure, pontifical mass was said with great solemnity. The king was there, and gave his last instructions orally to the young commander and presented him with a banner carrying the royal arms.

The ships were decked with many coloured flags. Musicians with their bagpipes, fifes, drums, and horns added to the liveliness. The people, both those who were to sail and those on shore, were dressed as for a fête. All Lisbon had come to see them off and to wish them good fortune, for this was the first commercial fleet to sail for India. The way had been found, and it was this voyage that was to bring back a rich reward in jewels and spices and pave the way for even greater wealth to follow.

On the following day, Monday the 9th of March 1500, the fleet left the mouth of the Tagus and departed from the Bay of Cascaes. All sails were set, and on them was displayed the red cross of the Order of Christ, for Cabral's fleet also went to bring the true Faith to the people of India. The conversion of the heathen was not only the desire of Dom Manuel, but it was an obligation imposed by the Pope. The bulls of 1493, granted by Alexander VI as the head of the Church, had given spheres of influence over non-Christian countries with the implied duty of bringing them under the guidance of Rome. This was shown in the bull addressed to Dom Manuel in June 1497. In it the Pope granted the request of the king and permitted him to possess the lands conquered from the infidels, provided no other Christian kings had rights to them, and prohibited all other rulers from molesting him. At the end he requested him to endeavour to establish the dominion of the Christian religion in the lands that he might conquer. This may explain the religious tone of the king's letter to the Zamorin of Calicut.

The voyage thus begun was to be the longest in history up to that time. The fleet sailed with the steady north-east trade wind and a favourable current over a course well known to the pilots, who had followed it many times to the coast of Guinea. On the following Saturday, the 14th March, they passed in sight of the Canary Islands and on Sunday the 22nd they reached São Nicolau of the Cape Verde Islands. No stop was made here, since it was not felt that supplies were needed. At daybreak the next morning the ship of Vasco de Ataíde was missing. The fleet searched for it for two days, but it was not found. Caminha, our most reliable authority, states that the weather was clear, although other authorities have claimed that there was a storm. This ship probably did not return to Lisbon, but the contemporary writers are at variance on this point. The fleet then continued its course, taking advantage of the north-east trade winds, and, in the hope of rounding the doldrums and the south-east trades, steered somewhat to the west of south. As the ships proceeded, the currents carried them further west.

We do not know the exact course followed, but apparently the equator was crossed at about the thirtieth meridian. The fleet then resumed its route to the south-south-west and followed the coast of Brazil at a distance because of better sailing conditions until Tuesday the 21st of April, when signs of land were encountered. The fleet continued its course, and Mount Pascoal on the coast of Brazil was sighted the next day.

Soon after land was discovered, the fleet cast anchor. On Thursday the 23rd of April, the smaller vessels went directly towards shore and a landing was made. Many of the natives were seen on the beach as soon as the boat neared the land. This was probably the first time the Portuguese had set foot on American soil within their sphere. The new land was named Terra da Vera Cruz.

On the shore they encountered strange people with bodies painted and tattooed, and decorated with coverings of brilliant feathers. They had dark complexions and long straight hair. Their appearance and customs were entirely unlike any that the Portuguese had seen before. Those who had landed immediately returned to the Chief Captain and reported that the port appeared to be a safe anchorage. When the fleet had cast anchor the boat again went ashore in order to obtain closer acquaintance with the inhabitants. These, however, did not await the near approach of the newcomers but fled and could not be prevailed upon to return either by means of signs or of gifts thrown upon the ground. A third attempt to approach them was equally resultless.

It had been the intention of the Chief Captain to land on the following day at this place but during the night the wind freshened to a gale - to the violence of which the ships were exposed and it was necessary to weigh anchor and to seek a more sheltered anchorage.

Late in the afternoon of the day after arrival in American waters the fleet arrived at a large bay that the Chief Captain entered, sounding with the lead. Here was found a safe haven sheltered from the gale and good holding-ground. We cannot suppose that the progress made on the day on which land was sighted or on the following day exceeded the average of one hundred and twenty-four miles maintained between Lisbon and Santiago. One hundred miles might, I think, be reasonably assumed to be accomplished each day. Such a rate of progress would result in the attainment of the vicinity of All Saints' Bay on the day after arrival. It is, moreover, very improbable that a bay that affords such good shelter and space for a large fleet to anchor in would be passed by. These considerations lead me to believe that All Saints' Bay (Bahia dos Todos Santos) was the Porto Seguro of Cabral.

At the second anchorage Nicolao Coelho was sent on shore and attempted to have speech with the inhabitants. These were of more confiding disposition than those who had been encountered at the first anchorage. They awaited the approach of the strangers and replied to the signs made to them. In complexion they were similar to those previously seen. Correa styles them a white people and adds that their noses resembled those of Javans. They were armed with bows of great length and arrows having arrow-heads of cane. Some sailors who went a few miles inland found that their villages consisted of wooden houses thatched with grass. They slept in nets suspended by the extremities (that we now call hammocks) and a few of them wore garments or cloaks made of cotton twist (fio d'algodão) to which brightly coloured feathers were attached. If this bay was Bahia dos Todos Santos it is probable that the inhabitants thus described were either of the Tupuia or of the Tupinamba tribe. Osorio relates that the Chief Captain had a stone monument or “padrão” erected similar to those that Vasco da Gama had set up.

The fleet remained here until the 2nd of May, trading with the natives and replenishing its supply of water and wood. No effort was made to explore the coast, and it was not learned whether it was an island or the mainland.

Pedro Cabral and the council of captains, which included Bartolomeu Dias, held it to be their duty to inform Dom Manuel speedily of the discovery that they had made. Pedro Cabral sent back to Portugal a supply-ship under the command of Gaspar de Lemos. In it there were several natives of the country as was customary at that period and also specimens of their handiwork such as feather cloaks and hammocks. Parrots and brazil-wood were also sent to Portugal. This supply-ship probably returned directly to Lisbon and did not follow the coast, as claimed by the Portuguese historian Gaspar Corrêa. There is no record of the date of its arrival in Portugal.

The supply-ship carried letters to inform the king of the new discovery. Two of these letters have been preserved, one written by Pero Vaz de Caminha, who tells of what occurred while they were on shore, and another, which is of a more scientific nature, by Master John, an astronomer.

CHAPTER 17

PERO VAZ DE CAMINHA'S ACCOUNT OF THE DISCOVERY

Pero Vaz de Caminha had accepted the position of writer in the fleet under Pedro Cabral's command. He sailed in Pedro Cabral's flagship with other writers. One of the two letters sent by Pedro Cabral to the King of Portugal relating their discovery of Brazil was written by him on 1st May 1500. It is the first document describing the discovery of Brazil, and has sometimes been called the first page in the history of Brazil. It was kept in the Torre do Tombo, and classified as Corpo Chronologico, gaveta 8, maco 2, no. 8.

****

"Senhor : Although the chief captain of this your fleet, and also the other captains, are writing to Your Highness the news of the finding of this your new land which was now found in this navigation, I shall not refrain from also giving my account of this to Your Highness, as best I can, although I know less than all of the others how to relate and tell it well. Nevertheless, may Your Highness take my ignorance for good intention, and believe that I shall not set down here anything more than I saw and thought, either to beautify or to make it less attractive. I shall not give account here to Your Highness of the ship's company and its daily runs, because I shall not know how to do it, and the pilots must have this in their charge.

And therefore, Senhor, I begin what I have to relate and say that the departure from Belem, as Your Highness knows, was on Monday, the 9th of March, and on Saturday, the 14th of the said month, between eight and nine o'clock, we found ourselves among the Canary Islands, nearest to Grand Canary; and there we remained all that day in a calm, in sight of them, at a distance of about three or four leagues. On Sunday, the 22nd of the said month, at ten o'clock, a little more or less, we came in sight of the Cape Verde Islands, that is to say, of the island of Sam Nicolao, according to the assertion of Pero Escolar, the pilot. On the following night, on Monday at daybreak, Vasco d'Atayde with his ship was lost from the fleet without there being there heavy weather or contrary winds to account for it. The captain used all diligence to find him, seeking everywhere, but he did not appear again. And so we followed our route over this sea until Tuesday of the octave of Easter, which was the 21st of April, when we came upon some signs of land, being then distant from the said island, as the pilots said, some six hundred and sixty or six hundred and seventy leagues; these signs were a great quantity of long weeds, which mariners call botelho, and others as well which they also call rabo de asno. And on the following Wednesday, in the morning, we met with birds they call fura buchos. On this day at the vesper hours we caught sight of land, that is, first of a large mountain, very high and round, and of other lower lands to the south of it, and of flat land, with great groves of trees. To this high mountain the captain gave the name of Monte Pascoal, and to the land, Terra da Vera Cruz. He ordered the lead to be thrown. They found twenty-five fathoms; and at sunset, some six leagues from the land, we cast anchor in nineteen fathoms, a clean anchorage. There we remained all that night, and on Thursday morning we made sail and steered straight to the land, with the small ships going in front, in 17, 16, 15, 14, 13, 12, 10, and 9 fathoms, until half a league from the shore, where we all cast anchor in front of the mouth of a river. And we arrived at this anchorage at ten o'clock, more or less. And from there we caught sight of men who were going along the shore, some seven or eight, as those on the small ships said, because they arrived there first. We there launched the boats and skiffs, and immediately all the captains of the ships came to this ship of the chief captain, and there they talked. And the captain sent Nicolao Coelho on shore in a boat to see that river. And as soon as he began to go thither men assembled on the shore, by twos and threes, so that when the boat reached the mouth of the river eighteen or twenty men were already there. They were dark, and entirely naked, without anything to cover their shame. They carried in their hands bows with their arrows. All came boldly towards the boat, and Nicolao Coelho made a sign to them that they should lay down their bows, and they laid them down. He could not have any speech with them there, nor understanding that might be profitable, because of the breaking of the sea on the shore. He gave them only a red cap and a cap of linen, which he was wearing on his head, and a black hat. And one of them gave him a hat of long bird feathers with a little tuft of red and grey feathers like those of a parrot. And another gave him a large string of very small white beads that look like seed pearls; these articles I believe the captain is sending to Your Highness. And with this he returned to the ships because it was late and he could have no further speech with them on account of the sea. On the following night it blew so hard from the south-east with showers that it made the ships drift, especially the flagship.

And on Friday morning, at eight o'clock, a little more or less, on the advice of the pilots, the captain ordered the anchors to be raised and to set sail. And we went northward along the coast with the boats and skiffs tied to the poop, to see whether we could find some shelter and good anchorage where we might lie, to take on water and wood, not because we were in need of them then, but to provide ourselves here. And when we set sail there were already some sixty or seventy men on the shore, sitting near the river, who had gathered there little by little. We continued along the coast and the captain ordered the small vessels to go in closer to the land, and to strike sail if they found a secure anchorage for the ships. And when we were some ten leagues along the coast from where we had raised anchor, the small vessels found a reef within which was a harbour, very good and secure with a very wide entrance. And they went in and lowered their sails. And gradually the ships arrived after them, and a little before sunset they also struck sail about a league from the reef, and anchored in eleven fathoms. And by the captain's order our pilot, Affonso Lopez, who was in one of those small vessels and was an alert and dextrous man for this, straightway entered the skiff to take soundings in the harbour. And he captured two well-built natives who were in a canoe. One of them was carrying a bow and six or seven arrows and many others went about on the shore with bows and arrows and they did not use them. Then, since it was already night, he took the two men to the flagship, where they were received with much pleasure and festivity.

In appearance they are dark, somewhat reddish, with good faces and good noses, well shaped. They go naked, without any covering; neither do they pay more attention to concealing or exposing their shame than they do to showing their faces, and in this respect they are very innocent. Both had their lower lips bored and in them were placed pieces of white bone, the length of a handbreadth, and the thickness of a cotton spindle and as sharp as an awl at the end. They put them through the inner part of the lip, and that part that remains between the lip and the teeth is shaped like a rook in chess. And they carry it there enclosed in such a manner that it does not hurt them, nor does it embarrass them in speaking, eating, or drinking. Their hair is smooth, and they were shorn, with the hair cut higher than above a comb of good size, and shaved to above the ears. And one of them was wearing below the opening, from temple to temple towards the back, a sort of wig of yellow birds' feathers, which must have been the length of a couto, very thick and very tight, and it covered the back of the head and the ears. This was glued to his hair, feather by feather, with a material as soft as wax, but it was not wax. Thus the head-dress was very round and very close and very equal, so that it was not necessary to remove it when they washed.

When they came on board, the captain, well dressed, with a very large collar of gold around his neck, was seated in a chair, with a carpet at his feet as a platform. And Sancho de Tovar and Simam de Miranda and Nicolao Coelho and Aires Correa and the rest of us who were in the ship with him were seated on the floor on this carpet. Torches were lighted and they entered, and made no sign of courtesy or of speaking to the captain or to any one, but one of them caught sight of the captain's collar, and began to point with his hand towards the land and then to the collar, as though he were telling us that there was gold in the land. And he also saw a silver candlestick, and in the same manner he made a sign towards the land and then towards the candlestick, as though there were silver also. They showed them a grey parrot that the captain brought here; they at once took it into their hands and pointed towards the land, as though they were found there. They showed them a sheep, but they paid no attention to it. They showed them a hen; they were almost afraid of it, and did not want to touch it; and afterwards they took it as though frightened. Then food was given them; bread and boiled fish, comfits, little cakes, honey, and dried figs. They would eat scarcely anything of that, and if they did taste some things they threw them out. Wine was brought them in a cup; they put a little to their mouths, and did not like it at all, nor did they want any more. Water was brought them in a jar; they took a mouthful of it, and did not drink it; they only washed their mouths and spat it out. One of them saw some white rosary beads; he made a motion that they should give them to him, and he played much with them, and put them around his neck; and then he took them off and wrapped them around his arm. He made a sign towards the land and then to the beads and to the collar of the captain, as if to say that they would give gold for that. We interpreted this so, because we wished to, but if he meant that he would take the beads and also the collar, we did not wish to understand because we did not intend to give it to him. And afterwards he returned the beads to the one who gave them to him. And then they stretched themselves out on their backs on the carpet to sleep without taking any care to cover their privy parts, which were not circumcised, and the hair on them was well shaved and arranged. The captain ordered pillows to be put under the head of each one, and he with the head-dress took sufficient pains not to disarrange it. A mantle was thrown over them, and they permitted it and lay at rest and slept.

On Saturday morning the captain ordered sails to be set and we went to seek the entrance, which was very wide and deep, six or seven fathoms, and all the ships entered within and anchored in five or six fathoms; this anchorage inside is so large and so beautiful and so secure that more than two hundred large and small ships could lie within it. And as soon as the ships were in place and anchored all the captains came to this ship of the chief captain, and from here the captain ordered Nicolao Coelho and Bartolameu Dias to go on shore, and they took those two men, and let them go with their bows and arrows. To each of them he ordered new shirts and red hats and two rosaries of white bone beads to be given and they carried them on their arms, with rattles and bells. And he sent with them to remain there a young convict, named Affonso Ribeiro, the servant of Dom Joham Tello, to stay with them, and learn their manner of living and customs; and he ordered me to go with Nicolao Coelho. We went at once straight for the shore. At that place there assembled at once some two hundred men, all naked, and with bows and arrows in their hands. Those whom we were bringing made signs to them that they should draw back and put down their bows, and they put them down, and did not draw back much. It is enough to say that they put down their bows. And then those whom we brought, and the young convict with them, got out. As soon as they were out they did not stop again, nor did one wait for the other; rather they ran, each as fast as he could. And they and many others with them passed a river which flows here with sweet and abundant water that came up as far as their waists. And thus they went running on the other side of the river between some clumps of palms, where were others, and there they stopped. And there, too, the young convict went with a man who, immediately upon his leaving the boat, befriended him, and took him thither. And then they brought him back to us, and with him came the others whom we had brought. These were now naked and without caps. And then many began to arrive, and entered into the boats from the seashore, until no more could get in. And they carried water gourds and took some kegs that we brought and filled them with water and carried them to the boats. They did not actually enter the boat, but from near by, threw them in by hand and we took them, and they asked us to give them something.

Nicolao Coelho had brought bells and bracelets and to some he gave a bell and to others a bracelet, so that with that inducement they almost wished to help us. They gave us some of those bows and arrows for hats and linen caps, and for whatever we were willing to give them. From thence the other two youths departed and we never saw them again.

Many of them, or perhaps the greater number of those who were there, wore those beaks of bone in their lips, and some, who were without them, had their lips pierced, and in the holes they carried wooden plugs that looked like stoppers of bottles. And some of them carried three of those beaks, namely, one in the middle and two at the ends. And others were there whose bodies were quartered in colour, that is, half of them in their own colour, and half in a bluish-black dye, and others quartered in checkered pattern. There were among them three or four girls, very young and very pretty, with very dark hair, long over the shoulders, and their privy parts so high, so closed, and so free from hair that we felt no shame in looking at them very well. Then for the time there was no more speech or understanding with them, because their barbarity was so great that no one could either be understood or heard. We made signs for them to leave, and they did so, and went to the other side of the river. And three or four of our men left the boats and filled I do not know how many kegs of water that we carried, and we returned to the ships. And upon seeing us thus, they made signs for us to return. We returned and they sent the convict and did not wish him to stay there with them. He carried a small basin and two or three red caps to give to their chief, if there was one. They did not care to take anything from him and thus they sent him back with everything, and then Bertolameu Dias made him return again to give those things to them, and he returned and gave them in our presence, to the one who had first befriended him. And then he came away and we took him with us. The man who befriended him was now well on in years, and was well decked with ornaments and covered with feathers stuck to his body, so that he looked pierced with arrows like Saint Sebastian. Others wore caps of yellow feathers, others of red, others of green; and one of the girls was all painted from head to foot with that paint, and she was so well built and so rounded and her lack of shame was so charming, that many women of our land seeing such attractions, would be ashamed that theirs were not like hers. None of them were circumcised, but all were as we were. And, thereupon, we returned, and they went away.

In the afternoon the chief captain set out in his boat with all of us and with the other captains of the ships in their boats to amuse ourselves in the bay near the shore. But no one went on land, because the captain did not wish it, although there was no one there; only he and all landed on a large island in the bay, which is very empty at low tide, but on all sides it is surrounded by water so that no one can go to it without a boat or by swimming. There he and the rest of us had a good time for an hour and a half, and the mariners fished there, going out with a net, and they caught a few small fish. And then, since it was already night, we returned to the ships.

On Low Sunday in the morning the captain determined to go to that island to hear mass and a sermon, and he ordered all the captains to assemble in the boats and to go with him; and so it was done. He ordered a large tent to be set up on the island and within it a very well-provided altar to be placed, and there with all the rest of us he had mass said, which the father, Frei Amrique, intoned and all the other fathers and priests who were there accompanied him with the same voice. That mass, in my opinion, was heard by all with much pleasure and devotion. The captain had there with him the banner of Christ, with which he left Belem, and it was kept raised on the Gospel side. After the mass was finished, the father removed his vestments, and sat down in a high chair, and we all threw ourselves down on that sand, and he preached a solemn and profitable sermon on the history of the Gospel, and at the end of it he dealt with our coming and with the discovery of this land, and referred to the sign of the Cross in obedience to which we came; which was very fitting, and which inspired much devotion.

While we were at mass and at the sermon, about the same number of people were on the shore as yesterday with their bows and arrows, who were amusing themselves and watching us; and they sat down, and when the mass was finished and we were seated for the sermon, many of them arose and blew a horn or trumpet and began to leap and to dance for a while, and some of them placed themselves in two or three almadias that they had there. These are not made like those I have already seen; they are simply three logs fastened together, and four or five, or all who wanted to, entered them, scarcely moving away at all from the land, but only far enough to keep their footing. After the sermon was finished the captain and all the rest proceeded to the boats with our banner displayed and we embarked, and thus we all went towards the land, to pass along it where they were, Bertolameu Dias going ahead in his skiff, at the captain's order, with a piece of timber from an almadia that the sea had carried to them, to give it to them. And all of us were about a stone's throw behind him. When they saw the skiff of Bertolameu Dias, all of them came at once to the water, going into it as far as they could. A sign was made to them to put down their bows, and many of them went at once to put them down on shore and others did not put them down. There was one there who spoke much to the others, telling them to go away, but they did not, in my opinion, have respect or fear of him. This one who was telling them to move carried his bow and arrows, and was painted with red paint on his breasts and shoulder blades and hips, thighs, and legs, all the way down, and the unpainted places such as the stomach and belly were of their own colour, and the paint was so red that the water did not wash away or remove it, but rather when he came out of the water he was redder. One of our men left the skiff of Bertolameu Dias and went among them, without their thinking for a moment of doing him harm; on the contrary, they gave him gourds of water and beckoned to those on the skiff to come on land. Thereupon Bertolameu Dias returned to the captain, and we came to the ships to eat, playing trumpets and pipes without troubling them further. And they again sat down on the shore and thus they remained for a while. On this island where we went to hear mass and the sermon the water ebbs a great deal and uncovers much sand and much gravel. While we were there some went to look for shell fish, but did not find them; they found some thick and short shrimps. Among them was a very large and very fat shrimp such as I had never seen before. They also found shells of cockles and mussels, but did not discover any whole piece. And as soon as we had eaten, all the captains came to this ship at the command of the chief captain and he went to one side with them and I was there too, and he asked all of us whether it seemed well to us to send news of the finding of this land to Your Highness by the supply ship, so that you might order it to be better reconnoitred, and learn more about it than we could now learn because we were going on our way. And among the many speeches that were made regarding the matter, it was said by all or by the greater number, that it would be very well to do so; and to this they agreed. And as soon as the decision was made, he asked further whether it would be well to take here by force two of these men to send to Your Highness and to leave here in their place two convicts. In this matter they agreed that it was not necessary to take men by force, since it was the general custom that those taken away by force to another place said that everything about which they are asked was there; and that these two convicts whom we should leave would give better and far better information about the land than would be given by those carried away by us, because they are people, whom no one understands nor would they learn quickly enough to be able to tell it as well as those others when Your Highness sends here, and that consequently we should not attempt to take any one away from here by force nor cause any scandal, but in order to tame and pacify them all the more, we should simply leave here the two convicts when we departed. And thus it was determined, since it appeared better to all.

When this was finished the captain ordered us to go to land in our boats in order to ascertain as well as possible what the river was like, and also to divert ourselves. We all went ashore in our boats, armed, and the banner with us. The natives went there along the shore to the mouth of the river, where we were going, and before we arrived, in accordance with the instructions they had received before, they all laid down their bows and made signs for us to land. And as soon as the boats had put their bows on shore, they all went immediately to the other side of the river, which is not wider than the throw of a short staff; and as soon as we disembarked some of our men crossed the river at once and went among them, and some waited and others withdrew, but the result was that we were all intermingled. They gave us some of their bows with their arrows in exchange for hats and linen caps and for anything else which we gave them. So many of our men went to the other side and mingled with them that they withdrew and went away and some went above to where others were. And then the captain had himself carried on the shoulders of two men and crossed the river and made every one return. The people who were there could not have been more than the usual number, and when the captain made all return, some of them came to him, not to recognise him for their lord, for it does not seem to me that they understand or have knowledge of this, but because our people were already passing to this side of the river. There they talked and brought many bows and beads of the kind already mentioned, and trafficked in anything in such manner that many bows, arrows, and beads were brought from there to the ships. And then the captain returned to this side of the river, and many men came to its bank. There you might have seen gallants painted with black and red, and with quarterings both on their bodies and on their legs, which certainly was pleasing in appearance. There were also among them four or five young women just as naked, who were not displeasing to the eye, among whom was one with her thigh from the knee to the hip and buttock all painted with that black paint and all the rest in her own colour; another had both knees and calves and ankles so painted, and her privy parts so nude and exposed with such innocence that there was not there any shame. There was also another young woman carrying an infant boy or girl tied at her breasts by a cloth of some sort so that only its little legs showed. But the legs of the mother and the rest of her were not concealed by any cloth.

And afterwards the captain moved up along the river, which flows continuously even with the shore, and there an old man was waiting who carried in his hand the oar of an almadia. When the captain reached him he spoke in our presence, without any one understanding him, nor did he understand us with reference to the things he was asked about, particularly gold, for we wished to know whether they had any in this land. This old man had his lip so bored that a large thumb could be thrust through the hole, and in the opening he carried a worthless green stone, which closed it on the outside. And the captain made him take it out; and I do not know what devil spoke to him, but he went with it to put it in the captain's mouth. We laughed a little at this and then the captain got tired and left him; and one of our men gave him an old hat for the stone, not because it was worth anything but to show. And afterwards the captain got it, I believe to send it with the other things to Your Highness. We went along there looking at the river, which has much and very good water. Along it are many palms, not very high, in which there are many good sprouts. We gathered and ate many of them. Then the captain turned towards the mouth of the river where we had disembarked, and on the other side of the river were many of them, dancing and diverting themselves before one another, without taking each other by the hand, and they did it well. Then Diogo Dias, who was revenue officer of Sacavem, crossed the river. He is an agreeable and pleasure-loving man, and he took with him one of our bagpipe players and his bagpipe, and began to dance among them, taking them by the hands, and they were delighted and laughed and accompanied him very well to the sound of the pipe. After they had danced he went along the level ground, making many light turns and a remarkable leap which astonished them, and they laughed and enjoyed themselves greatly. And although he reassured and flattered them a great deal with this, they soon became sullen like wild men and went away upstream. And then the captain crossed over the river with all of us, and we went along the shore, the boats going along close to land, and we came to a large lake of sweet water which is near the seashore, because all that shore is marshy above and the water flows out in many places. And after we had crossed the river some seven or eight of the natives joined our sailors who were retiring to the boats. And they took from there a shark which Bertolameu Dias killed and brought to them and threw on the shore. It suffices to say that up to this time, although they were somewhat tamed, a moment afterwards they became frightened like sparrows at a feeding-place. And no one dared to speak strongly to them for fear they might be more frightened; and everything was done to their liking in order to tame them thoroughly. To the old man with whom the captain spoke he gave a red cap; and in spite of all the talking that he did with him, and the cap which he gave him, as soon as he left and began to cross the river, he immediately became more cautious and would not return again to this side of it. The other two whom the captain had on the ships, and to whom he gave what has already been mentioned, did not appear again, from which I infer that they are bestial people and of very little knowledge; and for this reason they are so timid. Yet withal they are well cared for and very clean, and in this it seems to me that they are rather like birds or wild animals, to which the air gives better feathers and better hair than to tame ones. And their bodies are so clean and so fat and so beautiful that they could not be more so; and this causes me to presume that they have no houses or dwellings in which to gather, and the air in which they are brought up makes them so. Nor indeed have we up to this time seen any houses or anything that looks like them. The captain ordered the convict, Affonso Ribeiro, to go with them again, which he did. And he went there a good distance, and in the afternoon he returned, for they had made him come and were not willing to keep him there; and they had given him bows and arrows and had not taken from him anything that was his. On the contrary, he said, one of them had taken from him some yellow beads that he was wearing and fled with them; and he complained and the others at once went after him and returned to give them back to him. And then they ordered him to go back. He said that he had not seen there among them anything but some thatched huts of green branches, and made very large, like those of Entre Doiro e Minho. And thus we returned to the ships to sleep when it was already almost night.

On Monday after eating we all disembarked to take in water. Then many came there, but not so many as at the other times, and now they were carrying very few bows and they kept a little apart from us, and afterwards little by little mingled with us. And they embraced us and had a good time; and some of them soon slunk away. They gave there some bows for sheets of paper and for some worthless old cap, or for anything else. And in such a manner it came about that a good twenty or thirty of our people went with them to where many others of them were, with girls and women, and brought back many bows and caps of bird feathers, some green and some yellow, samples of which I believe the captain will send to Your Highness. And according to what those said who went there they made merry with them. On that day we saw them closer and more as we wished, for all of us were almost intermingled. And there some of them had those colours in quarters, others in halves, and others in such colours as in the tapestry of Arras, and all with their lips pierced, and many with the bones in them, and some of them without bones. Some of them were carrying prickly green nut shells from trees, that in colour resembled chestnuts, excepting that they were very much smaller. And these were full of small red grains that, when crushed between the fingers, made a very red paint with which they were painted. And the more they wetted themselves the redder they became. They are all shaved to above the ears, likewise their eyebrows and eyelashes. All of them have their foreheads from temple to temple painted with a black paint, that looks like a black ribbon the breadth of two fingers.

And the captain ordered that convict, Affonso Ribeiro, and two other convicts to go there among them, and likewise Diogo Dias, because he was a cheerful man, with whom they played. And he ordered the convicts to remain there that night. They all went there and mingled with them, and as they said later, they went a good league and a half to a village of houses in which there must have been nine or ten dwellings, each of which they said was as long as the captain's ship. And they were of wood with sides of boards and covered with straw, of reasonable height, and all had one single room without any divisions. They had within many posts, and from post to post a net is tied by the ends to each post, high up, where they sleep. And underneath they made their fires to warm themselves. And each house had two small doors, one at one end, and another at the other. And they said that thirty or forty persons dwelt in each house, and that thus they found them. And that they gave them to eat of the food that they had, namely, much manioc and other roots that are in the land, that they eat.

And, as it was late, they presently made all of us return and did not wish any one to remain there; and also, as they said, they wished to come with us. They traded there, for bells and for other trifles of little value that we were carrying, very large and beautiful red parrots, and two little green ones and caps of green feathers and a cloth of feathers of many colours, woven in a very beautiful fashion. All of these things Your Highness will see, because the captain will send them to you, as he says. And thereupon they came back and we returned to the ships.

On Tuesday, after eating, we landed to set a watch over the wood and to wash clothes. Some sixty or seventy men without bows or anything else were there on the shore when we reached it. As soon as we arrived they at once came to us without being frightened, and afterward many more came. There must have been a good two hundred, all without bows, and they all mingled so much with us that some of them helped us to load wood and put it in the boats, and they vied with us and derived much pleasure therefrom. And while we were taking on the wood two carpenters made a large cross from one piece of wood that was cut yesterday for this. Many of them came there to be with the carpenters; and I believe that they did this more to see the iron tools with which they were making it than to see the cross, because they have nothing of iron. And they cut their wood and boards with stones shaped like wedges put into a piece of wood, very well tied between two sticks, and in such a manner that they are strong, according to what the men said who were at their houses yesterday, for they saw them there. By now they kept us so much company as almost to disturb us in what we had to do. And the captain ordered the two convicts and Diogo Dias to go to the village, and to other villages if they should hear of them, and on no account to come to sleep on the ships, even if they should order them to; and so they went. While we were in this grove cutting wood, some parrots flew across these trees, some of them green, and others grey, large and small, so that it seems to me that there must be many in this land, but I did not see more than about nine or ten. We did not then see other birds except some pombas seixas, and they seemed to me considerably larger than those of Portugal. Some said that they saw turtle-doves, but I did not see any; but since the groves are so numerous and so large and of such infinite variety, I do not doubt that in the interior there are many birds. And towards night we returned to the ships with our wood. I believe, Senhor, that heretofore I have not given account to Your Highness of the form of their bows and arrows. The bows are black and long and the arrows long, and their tips of pointed reeds, as Your Highness will see from some which I believe the captain will send to you.

On Wednesday we did not go on shore, because the captain spent the whole day in the supply ship emptying it, and had transported to the ship what each one could carry. Many of the natives came to the shore, as we saw from the ships. There must have been some three hundred, according to what Sancho de Tovar said, who was there. Diogo Dias and Affonso Ribeiro, the convict, whom the captain sent yesterday to sleep there at any cost, returned when it was already night because they did not want them to sleep there, and they found green parrots and other birds that were black, almost like magpies, except that they had white beaks and short tails. And when Sancho de Tovar returned to the ship, some of them wished to go with him; but he did not want any except two proper youths. He ordered them to be well fed and cared for that night, and they ate all the food that was given them, and he ordered a bed with sheets to be made for them, as he said, and they slept and were comfortable that night. And so nothing more happened that day to write about.

On Thursday, the last of April, we ate early in the morning and went on shore for more wood and water, and when the captain was about to leave his ship Sancho de Tovar arrived with his two guests, and because he had not yet eaten, cloths were laid for him and food was brought, and he ate. We seated the guests in their chairs, and they ate very well of all which was given them, especially of cold boiled ham and rice. They did not give them wine, because Sancho de Tovar said that they did not drink it well. After the meal was over we all entered the boat and they with us. A sailor gave one of them a large tusk of a wild boar, well turned up. And as soon as he took it he at once put it in his lip; and because it did not fit there, they gave him a small piece of red wax. And this he applied to the back of his ornament to hold it and put it into his lip with the point turned upward, and he was as pleased with it as though he had a great jewel. And as soon as we disembarked he at once went off with it, and did not appear there again. When we landed there were probably eight or ten of the natives about, and little by little others began to come. And it seems to me that that day there came to the shore four hundred or four hundred and fifty men. Some of them carried bows and arrows and gave all for caps and for anything that we gave them. They ate with us of what we gave them. Some of them drank wine and others could not drink it, but it seems to me that if they accustomed themselves to it, they would drink it with great willingness. All were so well disposed and so well built and smart with their paints that they made a good show. They loaded as much of that wood as they could, very willingly, and carried it to the boats, and were quieter and more at ease among us than we were among them. The captain went with some of us for a short distance through this grove to a large stream of much water, which in our opinion was the same as the one that runs down to the shore, from which we took water. There we stayed for a while, drinking and amusing ourselves beside the river in this grove, that is so large and so thick and of such abundant foliage that one cannot describe it. In it there are many palms, from which we gathered many good sprouts. When we disembarked, the captain said that it would be well to go directly to the cross, which was leaning against a tree near the river, to be set up the next day, which was Friday, and that we should all kneel down and kiss it so that they might see the respect that we had for it. And thus we did. And we motioned to those ten or twelve who were there that they should do the same, and at once they all went to kiss it. They seem to me people of such innocence that, if one could understand them and they us, they would soon be Christians, because they do not have or understand any belief, as it appears. And therefore, if the convicts who are to remain here will learn their language well and understand them, I do not doubt that they will become Christians, in accordance with the pious intent of Your Highness, and that they will believe in our Holy Faith, to which may it please Our Lord to bring them. For it is certain this people is good and of pure simplicity, and there can easily be stamped upon them whatever belief we wish to give them; and furthermore, Our Lord gave them fine bodies and good faces as to good men; and He who brought us here, I believe, did not do so without purpose. And consequently, Your Highness, since you so much desire to increase the Holy Catholic Faith, ought to look after their salvation, and it will please God that, with little effort, this will be accomplished.

They do not till the soil or breed stock, nor is there ox or cow, or goat, or sheep, or hen, or any other domestic animal that is accustomed to live with men; nor do they eat anything except these manioc, of which there is much, and of the seeds and the fruits which the earth and the trees produce. Nevertheless, with this they are stronger and better fed than we are with all the wheat and vegetables that we eat.

While they were there that day, they continually skipped and danced with us to the sound of one of our tambours, in such a manner that they are much more our friends than we theirs. If one signed to them whether they wished to come to the ships, they at once made ready to do so, in such wise that had we wished to invite them all, they would all have come. However, we only took four or five this night to the ships, namely : the chief captain took two, and Simão de Miranda, one, whom he already had for his page, and Aires Gomes, another, also as a page. One of those whom the captain took was one of his guests whom we had brought him the first night when we arrived; today he came dressed in his shirt and with him his brother. These were this night very well entertained, both with food and with a bed with mattresses and sheets to tame them better.

And today, which is Friday, the first day of May, we went on land with our banner in the morning and disembarked up the river towards the south, where it seemed to us that it would be better to plant the cross, so that it might be better seen. And there the captain indicated where the hole should be made to plant it, and while they were making it, he with all the rest of us went to where the cross was down the river. We brought it from there with the friars and priests going ahead singing in the manner of a procession. There were already some of the natives there, about seventy or eighty, and when they saw us coming, some of them went to place themselves under it in order to help us. We crossed the river along the shore and went to place it where it was to be, which is probably a distance of two cross-bow shots from the river. While we were busy with this there came a good one hundred and fifty or more. After the cross was planted with the arms and device of Your Highness that we first nailed to it, we set up an altar at the foot of it. There the father, Frei Amrique, said mass, at which those already mentioned chanted and officiated. There were there with us some fifty or sixty natives, all kneeling as we were, and when it came to the Gospel and we all rose to our feet with hands lifted, they rose with us and lifted their hands, remaining thus until it was over. And then they again sat down as we did. And at the elevation of the Host when we knelt, they placed themselves as we were, with hands uplifted, and so quietly that I assure Your Highness that they gave us much edification. They stayed there with us until communion was over, and after the communion the friars and priests and the captain and some of the rest of us partook of communion. Some of them, because the sun was hot, arose while we were receiving communion and others remained as they were and stayed. One of them, a man of fifty or fifty-five years, stayed there with those who remained. While we were all thus he collected those who had remained and even called others. He went about among them and talked to them, pointing with his finger to the altar, and afterwards he lifted his finger towards Heaven as though he were telling them something good, and thus we understood it. After the mass was over the father took off his outer vestment and remained in his alb, and then he mounted a chair near the altar, and there he preached to us of the Gospel and of the apostles whose day this is, treating at the end of the sermon of this your holy and virtuous undertaking, which caused us more edification. Those who still remained for the sermon were looking at him, as we were doing. And the one of whom I speak called some to come there; some came and others departed. And when the sermon was over, Nicolao Coelho brought many tin crosses with crucifixes, which he still had from another voyage, and we thought it well to put one around the neck of each; for which purpose the father, Frei Amrique, seated himself at the foot of the cross, and there, one by one, he put around the neck of each his own tied to a string, first making him kiss it and raise his hands. Many came for this, and we did likewise to all. They must have been about forty or fifty. And after this was finished it was already a good hour after midday; we went to the ships to eat, and the captain took with him that same one who had pointed out to the others the altar and the sky, and his brother with him, to whom he did much honour. And he gave him a Moorish shirt, and to the other one a shirt such as the rest of us wore. And as it appears to me and to every one, these people in order to be wholly Christian lack nothing except to understand us, for whatever they saw us do, they did likewise; wherefore it appeared to all that they have no idolatry and no worship. And I well believe that, if Your Highness should send here some one who would go about more at leisure among them, that all will be turned to the desire of Your Highness. And if some one should come for this purpose, a priest should not fail to come also at once to baptise them, for by that time they will already have a greater knowledge of our faith through the two convicts who are remaining here among them. Both of these also partook of communion today. Among all those who came today there was only one young woman who stayed continuously at the mass, and she was given a cloth with which to cover herself, and we put it about her; but as she sat down she did not think to spread it much to cover herself. Thus, Senhor, the innocence of this people is such, that that of Adam could not have been greater in respect to shame. Now Your Highness may see whether people who live in such innocence will be converted or not if they are taught what pertains to their salvation. When this was over we went thus in their presence to kiss the cross, took leave of them, and came to eat.

I believe, Senhor, that with these two convicts who remain here, there stay also two seamen who tonight left this ship, fleeing to shore in a skiff. They have not come back and we believe that they remain here, because tomorrow, God willing, we take our departure from here.

It seems to me, Senhor, that this land from the promontory we see farthest south to another promontory that is to the north, of which we caught sight from this harbour, is so great that it will have some twenty or twenty-five leagues of coastline. Along the shore in some places it has great banks, some of them red, some white, and the land above is quite flat and covered with great forests. From point to point the entire shore is very flat and very beautiful. As for the interior, it appeared to us from the sea very large, for, as far as eye could reach, we could see only land and forests, a land that seemed very extensive to us. Up to now we are unable to learn that there is gold or silver in it, or anything of metal or iron; nor have we seen any, but the land itself has a very good climate, as cold and temperate as that of Entre Doiro e Minho, because in the present season we found it like that. Its waters are quite endless. So pleasing is it that if one cares to profit by it, everything will grow in it because of its waters. But the best profit that can be derived from it, it seems to me, will be to save this people, and this should be the chief seed that Your Highness should sow there. And if there were nothing more than to have here a stopping-place for this voyage to

Calicut, that would suffice, to say nothing of an opportunity to fulfil and do that which Your Highness so much desires, namely, the increase of our Holy Faith.

And in this manner, Senhor, I give here to Your Highness an account of what I saw in this land of yours, and if I have been somewhat lengthy you will pardon me, for the desire I had to tell you everything made me set it down thus in detail. And, Senhor, since it is certain that in this charge laid upon me as in any other thing which may be for your service, Your Highness will be very faithfully served by me, I ask of you that in order to do me a special favour you order my son-in-law, Jorge Do Soiro, to return from the island of Sam Thomé. This I shall take as a very great favour to me.

I kiss Your Highness's hands. From this Porto Seguro of your island of Vera Cruz today, Friday, the first day of May of 1500.

PERO VAAZ DE CAMINHA”

CHAPTER 18

THE ANONYMOUS NARRATIVE

One of the earliest and the most complete contemporary accounts of the voyage of Cabral was written by a member of the fleet. From what is known of the voyage of Cabral from other sources the Anonymous Narrative is substantially accurate.

It was written by somebody who took part in the voyage and who lived to return. It thus ranks second only to the letters of Pero Vaz de Caminha and Master John as an authoritative source. The author is unknown, but he was without doubt a Portuguese. From the careful and concise manner in which the account was written it appears to have been either an official record of the voyage or a narrative intended for publication.

Some clue as to the identity of the Portuguese author may be obtained from the narrative itself. He was on Pedro Cabral's ship or that of Simão de Miranda or Pedro de Ataíde after the storm, and he returned either with Pedro Cabral or Simão de Miranda. He seems to have been present when Pedro Cabral met the Zamorin and was on shore at the time of the massacre and was among those saved. Since Pedro Cabral was on board his ship during the uprising, he could not have been the author.

Only Frei Henrique, Nuno Leitao da Cunha, and a sailor are mentioned, of the twenty who escaped. It may have been the work of some nobleman who went with the fleet, but it seems more probable, from the careful manner with which it was written, that it was composed by someone whose duty it was to make this record, possibly one of the writers.

The only one holding this position whose name is known and who might have been the author is João de Sa, who had gone with Vasco da Gama as a writer and undoubtedly held a position of trust under Pedro Cabral. His duties would take him ashore at Calicut, and he returned with the fleet. Since Ravenstein considers that de Sa may have been the author of the so-called Roteiro of the voyage of Vasco da Gama, he may have had a similar duty to perform with Cabral's fleet.

The supplement to the narrative of the voyage shows an exactness in weights and values that indicates that it was made by one of the commercial men in the fleet, probably by either a factor or a writer, but not necessarily by the author of the Anonymous Narrative.

Some version of this account was known to the Portuguese historians who wrote at a later date, but no contemporary copy can now be found in Portugal. It seems to have reached Italy soon after the fleet returned.

Because of the interest taken in Cabral's voyage, this narrative was well known in Venice, for at least four early manuscripts still exist in the Venetian dialect, and it was included in the first edition of Paesi.

The following is that part of the Anonymous Narrative that deals with the discovery of Brazil.

****

"Wherein King Manuel in person consigned the royal standard to the captain.

In the year 1500 the Most Serene King of Portugal, called Don Manuel by name, sent his armada of ships, large and small, to the parts of India, in which armada there were twelve large and small ships. The captain-general of this armada was Pedro Aliares Cabrile, a fidalgo. These ships departed, both well equipped and in good order, with everything that they might need for a year and a half. Of these twelve ships, he ordered that ten should proceed to Calichut [Calicut] and the other two to Arabia, directing their course so that they might make a place called Zaffalle [Sofala] because they wished to establish trade with merchants in the said place, which place, Zaffalle, is found to be on the way to Calichut. In like manner the other ten ships carried merchandise that they might need for the said voyage. And on the 8th of the month of March of the said year they were ready, and on that day, which was Sunday, they went a distance of two miles from this city to a place called Rastello, where there is a church called Sancta Maria de Baller [Belem]. To this place the Most Serene King went in person to consign to the captain the royal standard for the said armada.

Monday, which was the 9th day of March, the said armada departed on its voyage with good weather.

On the 14th day of the said month the said armada passed the Island of Chanaria [Gran Canaria].

On the 22nd day it passed the Island of Capo Verde.

On the 23rd day one ship became separated from the said armada, so that no news of it has been heard from that day to this, nor can anything be learned of it.

How the ships ran because of the storm

On the 24th of April [actually the 22nd April], which was Wednesday of the octave of Easter, the aforesaid armada came in sight of land, with which they had great pleasure; and they went to it to see what land it was. They found it a land very abundant in trees, and there were people who were going there along the shore of the sea. And they cast anchor at the mouth of a small river. And after the said anchors were cast, the captain ordered a boat to be launched in the sea, in which he sent to see what people they were. And they found that they were people of dark colour, between white and black, and well built, with long hair. And they go nude as they were born, without any shame whatever, and each one of them carried his bow with arrows, as men who were in defence of the said river. On the aforesaid armada there was no one who understood their language. And having seen this, those in the boat returned to the captain; and then night came on. During that night there was a great storm.

On the morning of the following day we raised anchor, and in a great storm we skirted the coast towards the north (the wind was the sirocco) to see whether we might find some port where the aforesaid armada might stay. Finally we found a port where we cast anchor. There we found some natives who were fishing in their little barks. One of our boats went to where these men were and took two of them and these they brought to the captain to learn what people they were, and, as has been said, they did not understand one another either in speech or by signs. And that night the captain kept them with him. On the following day he ordered them to be dressed in shirts and coats and red caps. They were very content with this attire and marvelled at the things that were shown them. He afterwards ordered them to be put on shore.

A root from which they make bread, and their other customs

Likewise on that same day, which was the octave of Easter, the 26th day of April, the chief captain determined to hear mass, and he directed a tent to be set up in a place where he ordered an altar to be erected. And all those of the said armada went to hear mass and a sermon; whereupon many of those men joined them, dancing and singing, with their horns. And immediately after mass had been said they all left for their ships. The men of the land entered the sea as far as their armpits, singing and making merry and festivity. And then, after the captain had dined, the people of the said armada returned to land, taking solace and pleasure with those men of the land, and they began to trade with the men of the armada, and gave their bows and arrows for little bells and leaves of paper and pieces of cloth. Thus all that day our men took pleasure with them. And we found in that place a river of sweet water, and we returned late to the ships.

On the following day the chief captain decided to take in water and wood, and all those of the said armada went on shore. And the men of that place came to help them with the aforesaid wood and water, and some of our men went to the place where these men dwell, which was three miles away from the coast of the sea; and they bartered for parrots and a root called igname, which is their bread, which the Arabs eat. Those of the armada gave them bells and pieces of paper in payment for the said things. In this place we remained five or six days. In appearance these people are dark, and they go nude without shame, and their hair is long, and they pluck their beards. And their eyelids and over their eyebrows are painted with figures of white and black and blue and red. They have the lip of the mouth, that is, the lower lip, pierced. In the opening, they put a bone as large as a nail, and others wear there a long blue stone or a green one, and they hang from their lips. Women likewise go nude without shame and they are beautiful of body, with long hair. And their houses are of wood, covered with leaves and branches of trees, with many wooden columns. In the middle of the said houses and from the said columns to the wall they hang a net of cotton, that holds a man. And between the nets they make a fire. Thus in a single house there may be forty or fifty beds set up like looms.

Parrots in the newly discovered land

In this land we saw no iron nor any other metal. They cut wood with stone. And they have many birds of many sorts, especially parrots, of many colours; among them are some as large as hens; and there are other very beautiful birds. Of the feathers of the said birds they make the hats and caps that they wear. The land abounds with many kinds of trees and much and excellent water and ignames and cotton. In this place we did not see any animals. The land is large, and we do not know whether it is an island or mainland, but on account of its size we believe that it is terra firma. Its climate is very good. And these men have nets and are great fishermen and fish for various kinds of fish. Among these we saw a fish that they caught. It must have been as large as a barrel and longer and round, and it had a head like that of a pig and small eyes, and it had no teeth and had ears the length of an arm and the width of half an arm. Below its body it had two holes and the tail was an arm's length long and equally wide. It had no feet anywhere. It had hair like a pig and the hide was as thick as a finger, and its meat was white and fat like that of a pig. [This was undoubtedly a Manatee]. During these days that we stayed there, the captain determined to inform our Most Serene King of the finding of this land, and to leave in it two men, exiles, condemned to death, who were in the said armada for this purpose. And the said captain promptly dispatched a small supply ship that they had with them, in addition to the twelve ships aforesaid. This small ship carried the letters to the king. In these were contained what we had seen and discovered. After the said small ship was dispatched, the captain went on shore and ordered a very large cross to be made of wood, and he ordered it to be set up on the shore, and also, as has been said, left two convicts in the said place. They began to weep and the men of the land comforted them and showed that they pitied them.

CHAPTER 19

THE FINAL VOYAGE OF BARTOLOMEU DIAS

Barros asserts that Pedro Cabral's fleet set sail to resume the intended voyage on 3rd May 1500. If so, one might find therein the explanation of his bestowal of the name Terra da Santa Cruz upon the newly discovered country for the Festival of the Invention of the Holy Cross falls upon that day. Bartolomeu Dias was the captain of one of the ships.

When the fleet left the coast of Brazil it took its course to the Cape of Good Hope, with the evident intention of making a stop at São Bras. This sea had never been sailed previously, and this voyage from Brazil to the African coast may have been longer than any that had hitherto been made without sighting land.

The fleet continued with light winds. A comet was seen on the 12th of May that was in view for ten days. This was to the crew an ill omen.

The Anonymous Narrative had described these events : "The following day, which was the 2nd of May of the said year, the armada made sail on its way to go round the Cape of Good Hope. This voyage would be across the gulf of the sea, more than 1,200 leagues, that is, four miles to a league. On the 12th day of the said month, while on our course, there appeared a comet with a very long tail in the direction of Arabia. It was in view continuously for eight or ten nights."

On the 23rd of May, according to Barros and Castanheda, on the ninth of that month according to Osorio, as the fleet was proceeding on its way in a high sea and with the wind astern, the wind suddenly veered to the contrary direction and, before the sails could be lowered, four ships were overset by the violence of the wind and their crews thrown into the sea and drowned.

The Anonymous Narrative had described these events : "On Sunday, which was the 24th day of the said month of May, as all the armada was sailing together with a favourable wind, with the sails half set and without bonnets because of a rain that we had the day before, while we were thus sailing, there came on us a head wind so strong and so sudden that we knew nothing of it until the sails were across the masts. And at that moment four ships were lost with all on board, without our being able to give them aid in any way. The other seven ships that escaped were also almost lost. And thus we took the wind astern with the masts and sails broken. And we were at the mercy of God; and thus we went all that day. The sea was so swollen that it seemed that we were mounting up to the heavens. The wind changed suddenly, although the storm was still so great that we had no desire to set sails to the wind. And going thus with this storm, without sails, we lost sight of one another, so that the ship of the captain with two others took a different route. And another ship called Il Re with two others took another route, and the other one, alone, took still another. And thus we went twenty days through this storm without setting a sail to the wind."

One of the four ships was the ship of Bartolomeu Dias. He died by drowning off the same Cape of Good Hope he had distinguished himself by discovering. The captains of the other vessels were Symão de Pina, Gaspar de Lemos and Ayres Gomez de Silva. It was subsequently assumed that this disaster had occurred in the vicinity of certain islands but the assumption rested on a slender foundation.

On passing his padrão after making his discovery of the southern route around Africa "he took leave of it as from a beloved son whom he never expected to see again". Twelve years later, whilst sailing in Pedro Cabral's armada, Bartolomeu Dias perished almost within its sight. The site of his greatest discovery had become the site of his greatest tragedy.

BIBLIOGRAPHY

The following are the principle English language sources of information concerning Bartolomeu Dias. They also contain references to the original sources of information :

Bartolomeu Dias (Ernst Georg Ravenstein, William Brooks Greenlee, Pero Vaz de Caminha) [2010] - Biography of Bartolomeu Dias.

The voyages of Diogo Cão and Bartolomeu Dias 1482-88 (E.G. Ravenstein) [1986] - Descriptions of the voyages of Diogo Cão and the major voyage of discovery of Bartolomeu Dias.

Pedro Cabral (James Roxburgh MacClymont, William Brooks Greenlee, Pero Vaz de Caminha) [2009] - Biography of Pedro Cabral, who Bartolomeu Dias made his final voyage with.

Journal of the First Voyage of Vasco Da Gama, 1497-1499 (Alvaro Velho) [1898] - Contains all three contemporary descriptions of the first voyage of Vasco da Gama by Alvaro Velho, King Manoel and Girolamo Sernigi; three Portuguese accounts of Vasco da Gama's first voyage written in 1608, 1612, and 1646, plus details of the voyage.

VIARTIS

CPSIA information can be obtained at www.ICGtesting.com
Printed in the USA
BVOW03s1849141013

333715BV00012B/305/P